SCORPIONS

OF SOUTHERN AFRICA

JONATHAN LEEMING

To my parents Roy and Rachel Leeming and my brother Mark Leeming

Struik Publishers
(a division of New Holland Publishing
(South Africa) (Pty) Ltd)
Cornelis Struik House
80 McKenzie Street
Cape Town 8001

New Holland Publishing is a member of
Johnnic Communications Ltd.
Visit us at **www.struik.co.za**
Log on to our photographic website
www.imagesofafrica.co.za
for an African experience.

Log on to the author's southern African scorpions website
www.scorpions.co.za

First published in 2003
10 9 8 7 6 5 4 3 2

UROPLECTES INSIGNIS

PARABUTHUS SCHLECHTERI

The photographs on the title page and pages 23–25, 31, 34, 85, 86 were
kindly donated to the author by the Agricultural Research Council–PPRI.

Publishing manager: Pippa Parker
Managing editor: Helen de Villiers
Editor: Jeanne Hromník
Designer: Janice Evans
Cartographer & illustrator (pp 10, 15): David du Plessis

Reproduction by Hirt & Carter Cape (Pty) Ltd
Printed and bound by
Sing Cheong Printing Company

ISBN 1 86872 804 8

Front cover: *Opistophthalmus carinatus*
Back cover: *Parabuthus mossambicensis*
Title page: *Parabuthus granulatus*

PSEUDOLYCHAS PEGLERI

**Dr Lorenzo Prendini is kindly acknowledged for permitting the use
of his distribution maps as a reference for creating the maps in this book.**

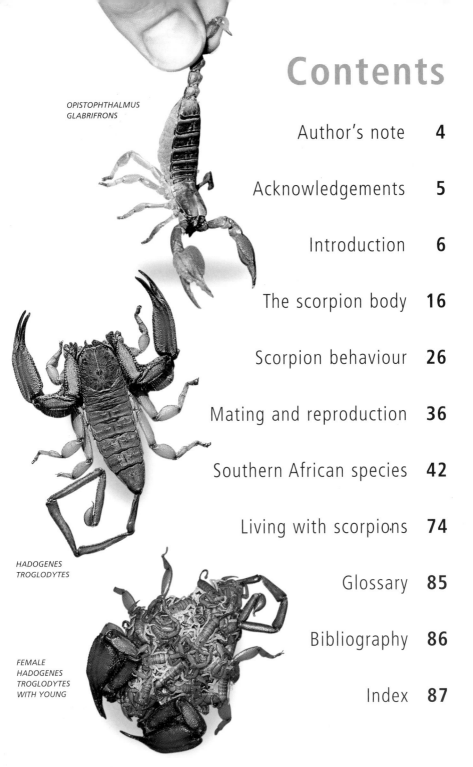

Contents

OPISTOPHTHALMUS
GLABRIFRONS

HADOGENES
TROGLODYTES

FEMALE
HADOGENES
TROGLODYTES
WITH YOUNG

Author's note

Although Africa is one of the best places on earth to study scorpions, little information on these much misunderstood and often persecuted creatures is available to the public. This book presents what I hope is an intriguing insight into the behaviour and ecology of southern African scorpions. It will help amateur naturalists as well as professionals in associated fields to identify our rich scorpion fauna and, I hope, will promote a better understanding and tolerance of scorpions.

On the whole, scorpions have a bad reputation. They are regarded as deadly venomous animals to be killed on sight, and it is forgotten that intricate life cycles, complex behavioural patterns and amazing adaptations to the environment lie beyond their apparent venomosity. This book aims to dispel myths about scorpions and replace them with facts about these fascinating creatures. It shows how to distinguish the more venomous from mildly venomous species. It describes how perfectly scorpions are adapted to their habitats and how the different parts of their anatomy are tailored to perform different functions.

Distribution maps are provided for all the species described in the species accounts in this book. They have been produced using museum and private collection records. These distributions are often biased towards well-collected areas such as the Kruger National Park in South Africa and desert regions in Namibia. Other areas such as Mozambique are under-sampled. The distribution maps in the book are intended as guides, not as absolutes.

The study of scorpions involves some very long words. I have tried to avoid complicated terminology, but this is not always possible. Meanwhile, the majority of family and genus names do not have common name equivalents and the different species must be identified using scientific names only. At the back of this book is a glossary that explains less familiar terms.

The classification of southern African scorpions is undergoing vigorous revision and contains more than a few grey areas of taxonomy. I have deliberately omitted species whose classification is unclear. The species included are those most likely to be encountered. In many cases, they have very wide distributions and are easily identified with the naked eye. Habitat is also a reliable guide when differentiating species in the wild. I have given practical information in the species accounts about where to look for scorpions.

Go out and take a look at the many small creatures around you. I have no doubt that after reading this book you will develop a new respect for scorpions and will view them from a fresh perspective.

Acknowledgements

This book is the result of the influence of many people, some of whom may be unaware of the part they have played. First of all, my parents, Rachel and Roy Leeming, who gave me their unlimited support and encouraged my unusual passion for scorpions, and Mark Leeming who grew up with a not so normal little brother.

My thanks go to Ian Engelbrecht for his invaluable comments and suggestions. Many of the photographs in this book are of specimens he has collected. I also wish to thank Martin Paulsen for his fantastic drawings; Jason Dunlop of the Berlin Museum for providing the pictures of the undescribed fossil scorpions and for his help with palaeoarachnology; Norman Larsen for photos and input; Moira Fitzpatrick for photos and advice on Zimbabwean species; Lorenzo Prendini for comments on the grey areas of scorpionology and for kindly making his distribution maps available as a source; Ansie Dippenaar-Schoeman for providing equipment from which the drawings were made and for the many years of encouragement.

Thanks also to Astri Leroy, Marie de Jager, Martin Filmer, Joh Henschel, Tanza Crouch and to the Jessnitz family for input and guidance with regard not only to the manuscript but all things creepy crawly. Many Spider Club members and private landowners have also contributed, many unknowingly.

Special thanks to Brad Segers and Dale Towert, two very special friends with whom I have shared many an adventure in search of scorpions. Whether it was cycling across the Namib Desert, scaling snowy mountain peaks or taking off on a wild goose chase in quest of that elusive scorpion species, thanks for being there.

My thanks to Pippa Parker and Jeanne Hromnik of Struik Publishers for encouragement and guidance and to Janice Evans for her design. And not least, thanks to all of you out there who have paused to look at scorpions and the other small creatures in our world.

Time for reflection in southern Namibia, home to a large number of southern Africa's scorpions.

INTRODUCTION

the
amazing
scorpion

SCORPIONS ARE TRULY amazing creatures. They inhabit tropical forests, rain forests, grasslands, savanna, temperate forests, caves over 800 m below the surface of the earth and snow-covered mountains more than 5 500 m in altitude. In some habitats they are the most successful predators in terms of density and diversity. They can survive extreme temperatures, from below freezing to above 50°C. Some species can be totally submerged under water for more than 48 hours without suffering any ill effects.

Of the 1 500 or so scorpion species known in the world, southern Africa has more than 130 species, widely distributed from the deserts of Namibia to the dune forests of KwaZulu-Natal, South Africa. They occur in every terrestrial habitat except the high mountain peaks of Lesotho. About six species occur on average in a localized area.

Scorpions are secretive creatures, preferring to hide their activities under the cloak of darkness. It is estimated that the average scorpion spends from 92 to 97 per cent of its time inactive. With such a high level of inertia, some species are able to live without food for more than a year.

Scorpions can be found in a variety of micro-habitats in temporary or permanent burrows under rocks, tree roots, loose bark and surface debris.

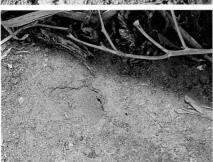

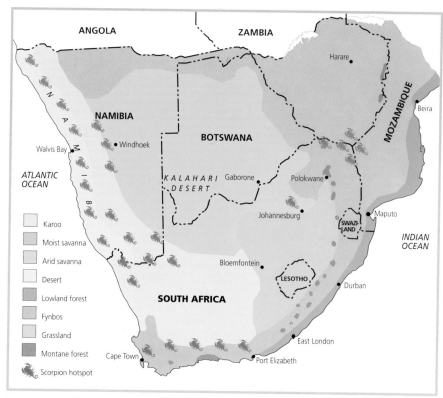

The southern African region, home to more than 130 of the world's scorpion species.

Although some species are small (20 mm long), others reach lengths of more than 210 mm and are among the largest terrestrial arthropods in the world. Life spans range from 2 to 10 years, some species living more than 30 years. Gestation periods range from 2 to 18 months, amazingly long for an arthropod when you consider that the gestation period of a humpback whale is just over 11 months. Excellent parental care ensures offspring mortality is kept low.

A scorpion's eyesight is poor, but it makes up for this with a stunning array of finely tuned senses. It uses sophisticated neural and behavioural mechanisms to sense prey vibrations. By accurately monitoring substrate vibrations, some species are able to capture prey burrowing below the ground. Other species have been known to capture flying prey by homing in on vibrations and wind currents that the prey's wings generate.

Of the described species of scorpions in the world, 25 are known to have caused human fatalities. Excluding bees and snakes, scorpions kill more people per year than any other nonparasitic animal. Statistics are scarce and often unreliable, but an estimated

5 000 people are known to die from scorpion stings every year. In Mexico alone, 1 696 deaths occur on average per year. India, North Africa, southern Africa and parts of South America are also inhabited by dangerously venomous scorpions.

In southern Africa, three species – *Parabuthus granulatus, P. transvaalicus* and (to a much lesser degree) *P. mossambicensis* – are responsible for the small number of fatalities recorded annually. *Hottentotta trilineatus* is medically important because of the strength of its venom, but has not been implicated in any deaths. It is probable that any large *Parabuthus* species can cause death in humans where there is evidence that the nervous system is already compromised. (See pages 34–35 and 76–77 for more on scorpion venom.)

OTHER ARACHNIDS

Scorpions belong to the class Arachnida and to the order Scorpiones. Their relatives in other arachnid orders include spiders (Aranaea), pseudoscorpions (Pseudoscorpiones), sunspiders (Solifugae), micro whipscorpions (Palpigradi), tailless whipscorpions (Amblypygi), harvestmen (Opiliones), mites and ticks (Acari), whipscorpions (Uropygi) and ricinuleids (Ricinulei). These creatures are some of the most misunderstood on earth.

All arachnids have eight segmented legs and exoskeletons that cover and protect their bodies. Many, but not all, are carnivorous. Arachnids use a variety of techniques to capture prey as not all have venom.

Among the major arachnid orders are:

SOLIFUGAE (Sunspiders): Although they look like their Aranaea relatives, solifugids do not have silk glands. They are covered in long hairs that glisten in the sun, have two simple eyes and, at first glance, appear to have 10 legs. The first pair are not legs at all, but pedipalps which have a sensory function, aid in feeding and have suckers at the tips. Solifugids are incredibly voracious. They prey upon almost any creature they can overpower, often much larger than themselves. Their appetites are insatiable and they gorge themselves on prey – spiders, scorpions, small vertebrates and insects – until they very nearly pop. It is ironic that such a fierce predator is so fragile.

Solifugids are distributed throughout southern Africa. They shelter in scrapes under surface debris and are often attracted to lights at night. They have extremely high metabolisms and rarely stand still even for a moment. An interesting fact is that they are known to collect the hair of humans and pets to make their nests. Solifugids do not have venom glands and are totally harmless to humans.

A seemingly fragile but ferocious solifugid predator.

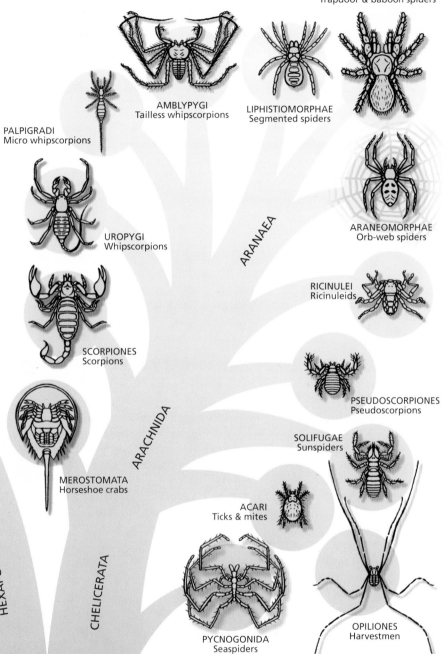

MYGALOMORPHAE
Trapdoor & baboon spiders

AMBLYPYGI
Tailless whipscorpions

LIPHISTIOMORPHAE
Segmented spiders

PALPIGRADI
Micro whipscorpions

ARANEAE

ARANEOMORPHAE
Orb-web spiders

UROPYGI
Whipscorpions

RICINULEI
Ricinuleids

SCORPIONES
Scorpions

PSEUDOSCORPIONES
Pseudoscorpions

ARACHNIDA

SOLIFUGAE
Sunspiders

MEROSTOMATA
Horseshoe crabs

ACARI
Ticks & mites

HEXAPODA

CHELICERATA

PYCNOGONIDA
Seaspiders

OPILIONES
Harvestmen

Arachnid orders of the class Arachnida, subphylum Chelicerata, phylum Arthropoda.

AMBLYPYGI (Tailless whipscorpions): Amblypygids are delicate looking and timid creatures with flattened bodies and extremely long legs (as much as 30 cm or longer in some parts of the world). Their bodies are ideally suited to life in narrow spaces. They appear to have a pair of whip-like feelers and six legs. In fact, the first pair of legs has been modified into 'feelers'. Their raptorial pedipalps are like those of

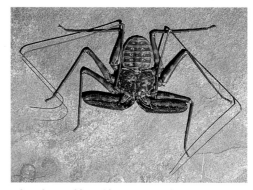

A harmless amblypygid, *Daemon variegatus.*

scorpions and are used to capture prey. They do not have venom glands and are totally harmless to humans.

Amblypygids are nocturnal and shelter in cracks in rocks, under the bark of trees or under surface debris during the day. They are often seen hunting around lights at night. They prey upon crickets, moths, millipedes and other small creatures, and can capture flying prey. They usually hunt close to their shelters and return to them at dawn. Very timid by nature, they escape predation by darting about with a sideways motion. If cornered, they stand on tiptoe and open their pedipalps in a threatening pose.

ARANAEA (Spiders): Along with scorpions, spiders are probably the most recognizable arachnids, occurring in almost every kind of terrestrial habitat and even some aquatic habitats. They occur in many different forms from ant mimics to spiders that look like bird droppings. Some are communal, others solitary. While some are huge, many are very small indeed. All except a single genus have venom glands. All spiders have silk glands, which they use to construct webs, nests and egg sacs. Some species construct webs to ensnare prey; others are active hunters. Spiders use their pedipalps in courtship, which is often a very complex affair.

Harpactira sp., a hairy member of the Aranaea.

HOW SCORPIONS DIFFER FROM OTHER ARACHNIDS

Scorpions are differentiated from other arachnid orders by their characteristic pincers and tail with a sting at the end. The pectines, which are comb-like structures on the underside of the scorpion, are also a distinguishing feature. All scorpions possess venom, but venom strength is very variable. In some species the venom is so weak that the scorpion is reluctant to use it to deter predators or catch prey.

Pincers and a sting-tipped tail characterize all scorpions.

THE QUESTION OF NAMES

The study of arachnids is difficult initially because of the terminology involved and the absence of common names for the majority of species. Common names for scorpions are few and are usually misleading. Scientific names must thus be used for reference.

Although common names have advantages, they differ from place to place and between different languages. Scientific names remain the same in any language (even Chinese and Russian) and usually describe a characteristic of the animal. They are also unique to each animal. The scientific naming system groups species in a hierarchy that shows clearly the relationships between similar animals. This system is strictly controlled by the Nomenclature Committee, an international organization which censors all animal and plant names.

Scientific names are more standardized than common names, but they do change over time. The scientific rules and regulations for naming a species stipulate that a species can have only one name. However, as research unveils new relationships between animals, flaws become apparent in the current classification. Two closely related species may become a single species, in which case the older name is used. It may also be necessary to split a single species into two distinct species. Changes can be made at all classification levels, not only at the level of species name.

When referring to a specific animal, only the genus and species names and not the whole classification – phylum, class, order, family, genus, species – are generally used. The genus name starts with a capital letter and is written in italics. The species name starts with a lower case letter and is also written in italics. For example: *Parabuthus transvaalicus*. In formal usage, the scientific name is followed by the name of the person who gave it and the year it was given. For example: *Parabuthus transvaalicus* (Purchell, 1899).

The brackets around the name and date indicate that the species was originally described by Purchell in 1899. This format is not necessary in common use.

If the author has used the genus name in full and refers to it again, the genus name may be abbreviated – as in *P. transvaalicus*. However, if another genus name starting with the same letter (e.g. *Pseudolychas pegleri*) is mentioned, *Parabuthus transvaalicus* is written out again in full. Reference to all animals of a genus is achieved by writing spp. after the genus name. The statement '*Parabuthus* spp. are highly venomous' is the same as 'All members of the genus *Parabuthus* are highly venomous'. (*P.* spp. is not acceptable.)

An unnamed species may be referred to by the genus name followed by sp. A subspecies is referred to as ssp. The current trend, however, is to do away with subspecies and to include the animal in another species or else to create a new species name.

The scientific scheme was invented by Carolus Linnaeus, a Swedish botanist born in 1707, who produced an extensive system of classification of plant and animal species. Because of limited knowledge at the time, groups were very large and contained distantly related species. Linnaeus's binomial system has been refined, but the basic principles are unchanged.

Sounds like Greek? That's because most of these names originate from Greek and Latin. It may be complicated initially, but it's very easy to work with once you get the idea.

Carolus Linnaeus, 1707–1778

WHAT'S IN A NAME?

A good example of an informative scientific name is *Brontoscorpius anglicus*, the name of an extinct scorpion. *Bronto* suggests a large animal (this scorpion was estimated to be over a metre in length); *scorpius* indicates the animal was a scorpion; *anglicus* refers to where it was discovered – in England. Therefore, the scientific name *Brontoscorpius anglicus* means 'giant scorpion from England'.

Sometimes the name refers to a locality, as in the case of *Parabuthus transvaalicus*, a member of the genus *Parabuthus* living in the former Transvaal region of South Africa. The name of the extinct family Gigantoscorpionidae conjures visions of a massive beast. In other instances, names refer to persons. *Hadogenes newlandsi*, for example, was named after Gerry Newlands in recognition of his work in southern African arachnology.

EARLY HISTORY

Scorpions are among the earliest forms of land animal, and their basic body plan has remained relatively unchanged for hundreds of millions of years. The earliest known fossil scorpions date back 425 to 450 million years – 200 million years before the first dinosaurs roamed the earth. Fossil scorpions of about 90 species have been discovered.

An exceptional array of behavioural, physiological and ecological adaptations has secured the success of scorpions for over 400 million years of geological and environmental change. Today, scorpions inhabit all major land masses except Antarctica. Man has introduced them to England and New Zealand.

Some scientists speculate that scorpions evolved from sea-living Eurypterida or waterscorpions. These early scorpions had gills, which were protected beneath abdominal plates, and large compound eyes. Their legs were not adapted to walking on land. Marine and amphibious scorpions appear to have co-existed about 350 to 400 million years ago.

The lifestyle of these early scorpions was similar to that of crabs today. They inhabited lagoons and estuaries, and scavenged and hunted along the shoreline and in shallow waters. At the time when these amphibious scorpions lived, arthropods such as millipedes and wingless insects had already colonized the land.

The most notable morphological changes in scorpion construction occurred during the transition from aquatic to terrestrial habitats. Since then scorpions have evolved without radical change. The first terrestrial scorpions appeared about 325 to 350 million years ago. The first totally terrestrial scorpion fossil dates back about 320 million years.

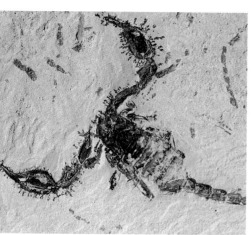

Fossilized scorpion species from Brazil – as yet unnamed.

The study of extinct scorpions is problematic because of the sparse fossil record. Many terrestrial scorpions lived in deserts or forests in which fossilization was a rare event. Although some species were very large indeed, many were relatively small and fragile creatures that were unsuitable for fossilization. Many fossils exist only as isolated fragments, rarely as whole preserved scorpions. Commonly, only the top (dorsal) or bottom (ventral) surface of the fossil is exposed, forcing palaeoarachnologists to describe and classify a specimen from a limited number of characteristics. *Brontoscorpius anglicus*, which dates back 380 million years, is known from only a single pedipalp finger – from which the scorpion is estimated to have been almost a metre long.

The length of early species ranged from 300 to 700 mm although some were over a metre in length. Larger scorpions benefited from an aquatic habitat, the water helping to support their bodies. This would have been especially important for species that are thought to have returned to the water to moult.

MYTHS AND LEGENDS

Scorpions have captured the attention of humans from very early times, occupying a prominent position in Egyptian, Greek and Chinese civilisations. Usually associated with witchcraft or evil, they are a source of myths in many cultures. The morbid interest they arouse originates no doubt from their venomous tendencies.

The most popular scorpion myth is that a scorpion will commit suicide when it is surrounded by a ring of fire. This is nonsense since scorpions are immune to their own venom. No animal except man avoids suffering by committing suicide, the suicidal instinct being of little practical advantage and unlikely to have survived the process of natural selection.

What actually happens is that the scorpion's co-ordination rapidly deteriorates with the heat of the flames. It defends itself by lashing out wildly with its tail. It may indeed sting itself during this time. Death results from the heat of the fire and not from a self-inflicted sting.

It is paradoxical that although scorpions have featured so prominently in ancient cultures, and arouse such deep-seated fears, we know so little about their natural history. Only now are we beginning to understand their complex behaviour, biology and physiology.

As in Greek mythology, Scorpio continues to chase Orion across the sky.

THE SCORPION BODY

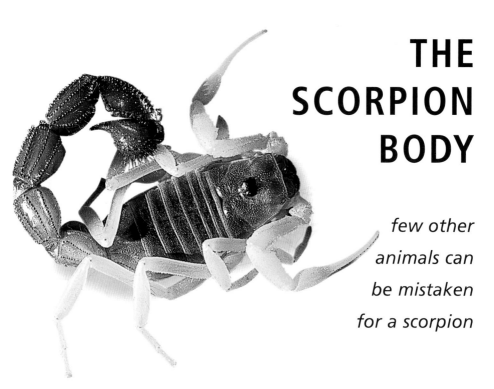

*few other
animals can
be mistaken
for a scorpion*

THE SCORPION BODY plan is easily recognized and is basically the same in all scorpions, although subtle variations have developed with the lifestyle of different groups. These differences contribute to the survival success rates of scorpions in different environments. Each facet of a scorpion's physical make-up has been honed over millions of years.

BODY PLAN

The body of a scorpion can be divided into three main regions: prosoma, mesosoma and metasoma. The metasoma (tail) is much segmented; the mesosoma shows external segmentation only. Together they are referred to as the opisthosoma.

Like other arthropods (invertebrates with jointed limbs and exoskeletons), including insects, scorpions have an exoskeleton made of numerous layers of chitin, a tough horny material. The exoskeleton protects the internal organs and supports the muscles. It also provides points of muscle attachment that, in turn, facilitate movement and locomotion.

The thick horny plates of chitin that form the exoskeleton or outer covering of the scorpion's body provide excellent protection but are unable to flex. For this reason, they are joined together by intersegmental membrane. This membrane provides flexibility and points of articulation for the thick protective plates. Being expandable, it allows the scorpion's body to temporarily increase and decrease in volume. The exoskeleton contains a waxy substance (known as lipids) that helps reduce water loss.

As a scorpion's body grows in volume, the exoskeleton becomes too small and is shed. This process is called ecdysis. The exact cues for ecdysis are not fully understood. In some cases, time of year plays a part; in others, body volume over a certain threshold triggers the process. Scorpions always undergo ecdysis in a protected area such as a burrow or crevice.

ECDYSIS

All animals with exoskeletons must undergo ecdysis – a process in which they shed their skins in order to grow larger. The new exoskeleton is secreted underneath the existing one, and absorbs some components of the old exoskeleton.

Approximately 24 hours prior to ecdysis, the individual becomes inactive. The old exoskeleton splits at the front and sides of the carapace. The scorpion draws itself out by brief vigorous movements followed by periods of rest. The pedipalps, legs, meso-

Opistacanthus leavipes shedding its skin.

soma and pectines are all drawn out of the old exoskeleton. This may take more than 12 hours.

After the scorpion has struggled free, it is very pale in colour and very vulnerable as it is unable to move while the exoskeleton is soft. It is also in danger of drying out. The periods between ecdysis are called instars. A newly born scorpion is in its first instar.

It is interesting that newly moulted scorpions do not fluoresce under ultraviolet light until the exoskeleton hardens. The old exoskeleton, however, fluoresces brightly. As the exoskeleton hardens, normal coloration returns. The average scorpion moults five or six times in its lifetime. In many species, males do not moult as many times as females.

Prosoma

The scorpion's prosoma (broadly speaking, its head) is covered by the carapace, a shield-like plate through which the eyes protrude. The scorpion's brain is situated underneath the carapace which, in some species, is characterized by a suture down the centre. On the underside of the prosoma are the pedipalps, legs, sternum and pectines.

Mesosoma

Connected to the prosoma is the mesosoma, which could be considered the body of the scorpion. It is protected by five thick shield-like plates on the underside called sternites and seven plates on top called tergites. The sternites have openings (spiracles) for the booklungs. The tergites are often keeled and may be used as key features in differentiating scorpion species.

Metasoma

Last (and not least) is the tail of the scorpion, which connects with the mesosoma and consists of five caudal segments, each progressively longer. The telson (sting) is at the end of the last segment. Venom is secreted by a pair of venom glands in the sting. It is stored in a reservoir in the venom vesicle in the sting. When the muscles around this reservoir contract, venom is squeezed out of the venom duct, which lies just behind the tip of the sting. In some species, venom can be replenished in a matter of hours. (For more on scorpion venom, see pages 34–35, 76–77.)

The sharp part of the venom vesicle is called the aculeus. In some species, a tooth-like projection (the subaculeur tubercle) can be seen on the inside curvature of the aculeus. The shape of the venom vesicle is often used as a diagnostic characteristic when identifying different species of scorpions.

The scorpion's tail cannot bend backwards, only forwards. It is used by *Parabuthus* species, for example, to dig burrows and by some species to produce sound.

Two tails: a *Hadogenes* with a weak and slender tail and small venom vesicle (left) and a *Uroplectes* (right) with a thick tail and large venom vesicle.

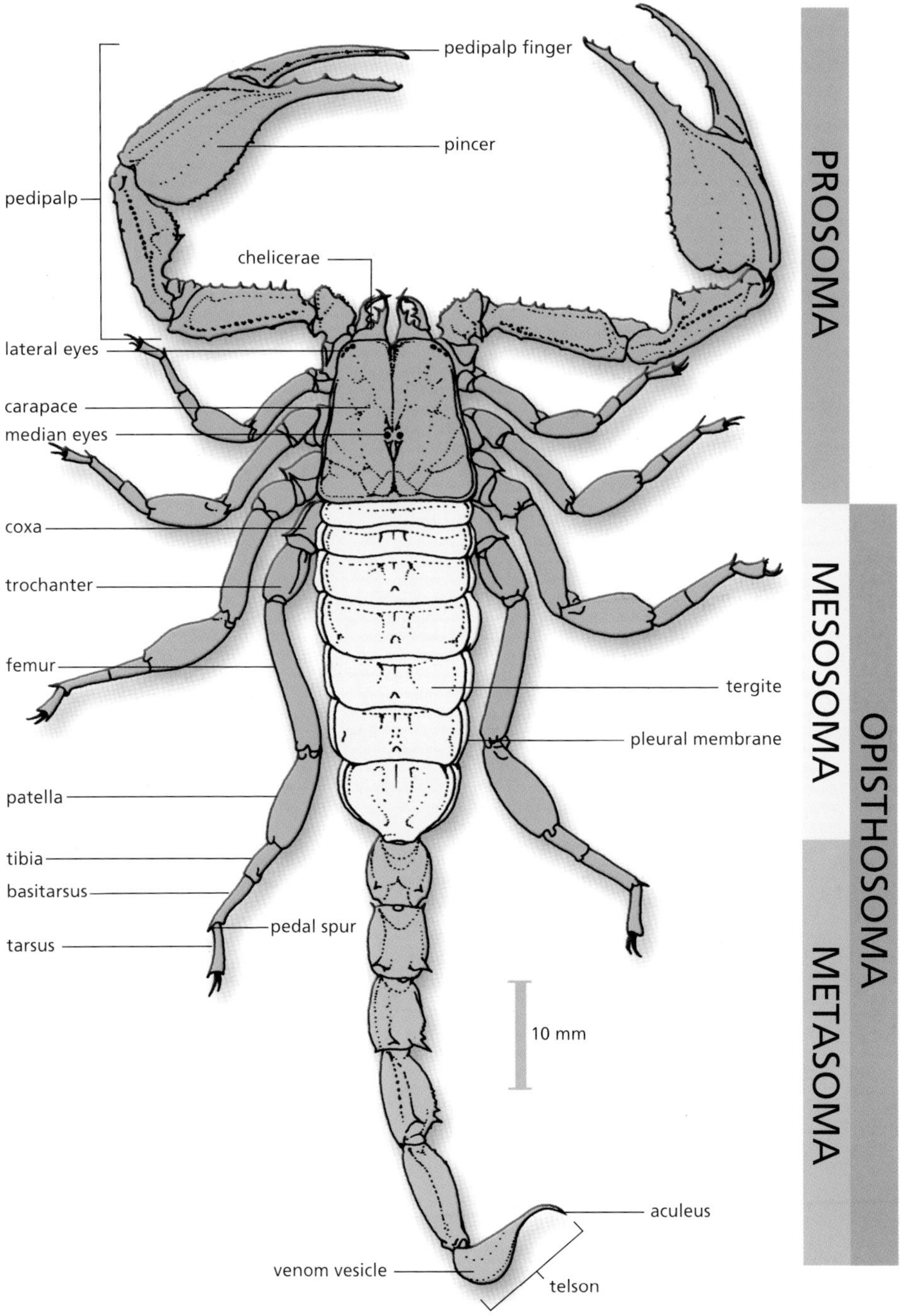

pedipalp finger

pincer

pedipalp

chelicerae

lateral eyes

carapace

median eyes

coxa

trochanter

femur

tergite

pleural membrane

patella

tibia

basitarsus

tarsus

pedal spur

10 mm

aculeus

venom vesicle

telson

PROSOMA

MESOSOMA

METASOMA

OPISTHOSOMA

THE SCORPION BODY

BODY PARTS

Eyes

All southern African scorpions have eyes, unlike cave-dwelling species in Mexico and Australia, which are highly adapted to their dark environment and have no eyes at all.

A few southern African species have six eyes (two median eyes and two groups of two lateral eyes), but most have eight (two median eyes and two groups of three lateral eyes located at the front corners of the carapace). The scorpion's two median eyes are situated on the prosoma often at the centre of the carapace and are usually about three times as large as the lateral eyes.

The eyes of scorpions are simple eyes, each consisting of a single lens that projects an image onto the nerves at the back of the eye, providing the brain with a single image. In compound eyes (for example, those of flies) each eye consists of numerous lenses, each lens generating an image. A scorpion's lateral eyes are very light-sensitive and can detect subtle differences in brightness. The median eyes are also very light-sensitive and are capable of discerning depth and forming rudimentary visual images.

Carapace of *Opistophthalmus longicauda* showing the large median eyes and – at the front corners – the much smaller lateral eyes.

Eyes are not the only light-sensing organs on a scorpion. Experiments have shown that some species can detect light using the tail. It is not known whether all scorpions possess this ability or how a scorpion can benefit from this feature.

Chelicerae

The chelicerae (mouthparts) are a pair of pincer-like appendages that are used to chew food before it enters the oral cavity behind them, which is where the digestive process starts. The chelicerae are capable of ripping prey apart. Scorpions also use their chelicerae during courtship and to groom themselves. When identifying scorpion species, the shape of the chelicerae is used as a diagnostic characteristic.

Scorpions are very clean creatures. They groom themselves with their chelicerae.

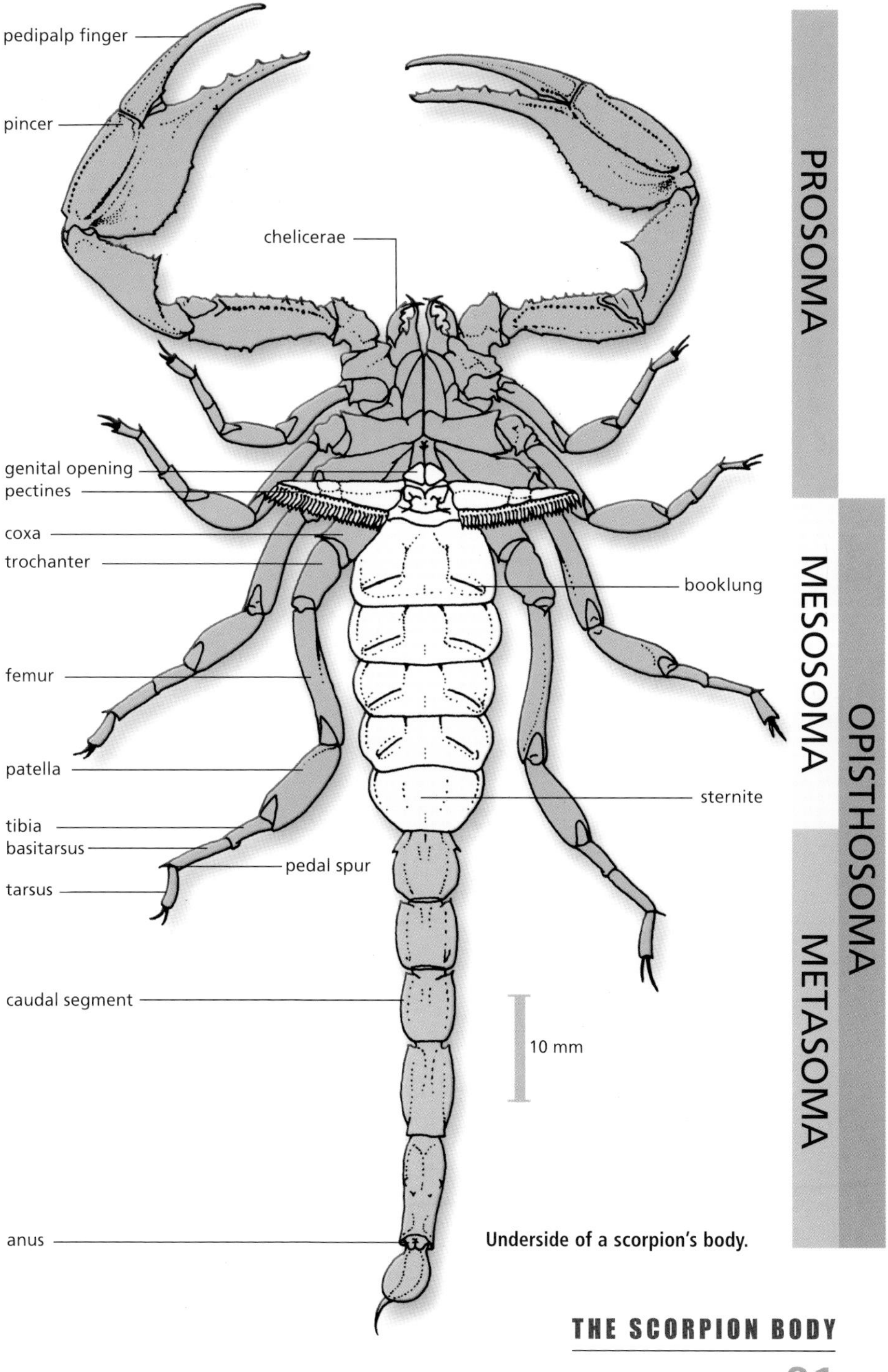

pedipalp finger

pincer

chelicerae

genital opening
pectines

coxa
trochanter

booklung

femur

patella

tibia
basitarsus

pedal spur

tarsus

sternite

caudal segment

10 mm

anus

Underside of a scorpion's body.

PROSOMA

MESOSOMA

OPISTHOSOMA

METASOMA

THE SCORPION BODY

Pedipalps

The scorpion's pedipalps are a pair of three-segmented appendages ending in the pincer. Although used primarily to grasp objects, pincers have a sensory function employed in courtship, defence and prey capture. When threatened, some species use their pincers as a shield. In many species, the pincers of males and females differ, the degree of difference depending on genus and, often, species.

Pincer (hand) of *Opistophthalmus karrooenis*.

Legs

Unlike insects, arachnids have four pairs of legs, eight in total. The legs consist of seven segments (coxa, trochanter, femur, patella, tibia, basitarsus and tarsus). The pedal spur, which is used for traction, can be found between the tarsus and basitarsus.

Because scorpions are ideally adapted to their environments, features such as their legs indicate preferred habitat. Scorpions that burrow have short and robust legs with stiff spines on them to aid in traction on hard substrate. Short legs are also an advantage when operating in a burrow. *Hadogenes* species possess strongly recurved claws at the end of their feet and their legs are flat in profile to suit their lifestyle. The claws on their feet allow them to walk inverted and to keep a tenacious grip on smooth rocky surfaces. In contrast, the feet of sand-living scorpions are characterized by numerous stiff hairs which make up sand combs that act like sand shoes on sandy substrate. These sand combs also aid in digging and shovelling away sand.

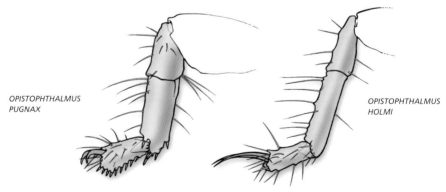

OPISTOPHTHALMUS
PUGNAX

OPISTOPHTHALMUS
HOLMI

Stiff hairs (setae) and highly recurved claws characterize the foot of the rock-dwelling species on the left. A sand-living scorpion of the same genus has extremely long tarsal claws (right) which aid in locomotion in soft sand, functioning like snow shoes.

Pectines

Just behind the sternum and genital opening are the pectines – comb-like structures on the underside of the prosoma that are unique to scorpions. Early arachnologists speculated that they were used for flying or as braces when the scorpion gave birth. It is now known that these structures have several different functions. They are used for sensing temperature and humidity levels and for feeling the substrate.

The pectines are composed of numerous pectinal teeth, the number varying within a single species and between sexes. Male scorpions almost always have more pectinal teeth than do females. In males of species such as *Uroplectes planimanus*, the first pectinal tooth is greatly elongated, and in *Pseudolychas pegleri* it is sickle-shaped. The presence of a modified pectinal tooth enables one to distinguish between different species.

The shape of the pectines is clearly visible on the shed skin of a *Parabuthus* species.

It is known that scorpions leave a scent trail behind them, the female giving off pheromones which the male senses. The pectines are where these scent trails are detected.

BODY FUNCTIONS

Circulation

A scorpion's heart is tube-like, consisting of several valved openings that extend down the entire length of the mesosoma. Scorpions do not have veins like humans. Instead, the internal organs float in hemolymph (the scorpion's 'blood') which fills the body cavity (hemoceol) of the scorpion

Hemolymph carries oxygen to the internal organs and muscles and transports waste products away from them. Seven valves in the heart allow the hemolymph to flow from the booklungs (where it is oxygenated) to the internal organs and muscles (where the oxygen is used). This hemolymph is then pumped back to the booklungs to be re-oxygenated and the cycle continues.

Metabolic waste in the hemolymph is absorbed by the malpighian tubes, which branch off the intestine at the end of the mesosoma, just before the metasoma.

Respiration

Scorpions do not have lungs like those of mammals. Instead, respiration is achieved through booklungs, so called because they look like pages in a book. All scorpions have four pairs of booklungs. Oxygen passes through the 'pages' (lamellae) of these 'lungs' into the hemolymph in the scorpion's body cavity.

The page-like lamellae – which may number between 140 and 150 – increase the absorptive capacity of the booklungs. They are held apart by bristles. Special muscles circulate hemolymph between the lamellae and force air in and out of the booklungs. A scorpion's oxygen requirements are very small in comparison with those of humans even if we consider the difference in mass. This enables some species to live inside a sealed-off burrow for many months.

Digestion

Unlike spiders, scorpions do not have fangs. They chew their food using their pincer-like chelicerae. Prey is initially chewed and mixed with saliva, which starts the digestive process. After the food is chewed and broken down into smaller pieces, it passes into the oral cavity, which is lined with numerous hairs and enzyme-producing cells. The hairs strain out any indigestible food; the cells produce enzymes that promote digestion.

Research has shown that some species of scorpion, at least, can taste their food. Their 'taste buds' are situated on their chelicerae.

Only predigested food passes from the oral cavity into the pharynx and then into the oesophagus and stomach. Digestion takes place in the stomach, which occupies most of the mesosoma and part of the prosoma. The digestive gland also provides a storage place for food reserves. Their ability to store large amounts of food, coupled with a slow metabolic rate, allows many scorpions to go without food for as long as a year or more. They may gain as much as a third of their body weight from a single feeding.

A scorpion's anus is situated at the end of its tail, between the last caudal segment and the venom vesicle.

Because it is strained through hairs in the oral cavity, a scorpion's food contains very little indigestible matter. As a result, scorpions produce very little fecal matter and lose very little water. The fecal matter of scorpions is liquid-like and white in colour.

Nervous system

The scorpion's neural ganglion (brain) is found under its median eyes. The ventral nerve runs down the length of the body, connecting the neural ganglion with the scorpion's highly developed sensory organs, which are richly supplied with nerves.

Scorpions have an acute and highly evolved sense of 'touch'. The microscopic slit sense organs on the legs and tail contort when contact and cuticular stress is exerted on the exoskeleton. The degree of contortion provides a highly sensitive measure of stress. Scorpions can detect ground vibrations in the substrate through the sense organs on their tarsi (feet) and, furthermore, can assess from which direction the vibrations are coming and from what distance.

In addition, they possess setae (hairs) on various parts of the exoskeleton. Many of these hairs have sensory functions and respond to chemical and physical signals. Stiff hairs on the tarsi can sense substrate vibrations up to 15 cm away. Setae may also be positioned on the body to suit the environment in which the scorpion lives. They may aid in digging or provide 'snow shoes' for walking on sand or for extra grip on rocky surfaces.

Trichobothria are specialized elongated (0,5–1,5 mm) sensory hairs situated in pits in the exoskeleton. In some species, these pits are visible to the naked eye. Smaller hairs line the interior of the pit. Trichobothria are used mainly to detect air currents and are arranged in distinctly different patterns, depending on the species. These patterns help to identify scorpion species.

Sensory devices such as hairs on the pedipalps compensate for a scorpion's poor eyesight.

SCORPION BEHAVIOUR

for any animal living in a harsh environment, there are many challenges to overcome

MANY SCORPIONS LIVE in environments where rapid water loss, exposure to ultraviolet light, scarcity of prey and environmental change are experienced regularly. They have a relatively large body surface area in relation to body mass and can thus experience rapid rates of water loss and heat gain. To meet the challenges of the unfavourable conditions in which they often live, scorpions display behavioural strategies as fascinating as their physical adaptations.

STRATEGIES FOR SURVIVAL

Scorpions are normally solitary creatures. No social scorpions occur in southern Africa, although *Uroplectes lineatus* has been found sheltering in groups under rocks on and around Table Mountain in the winter months (possibly because of a shortage of over-wintering sites). An exception to the rule, *Opistacanthus capyporum* from South America, lives in groups inside termite mounds, where each scorpion has its own chamber inside the mound. These scorpions co-operate in hunting prey.

It is estimated that the average scorpion is inactive almost all the time. Scorpions exhibit very low metabolic rates, and can escape prolonged periods of drought or harsh cold by simply waiting in their shelters for conditions to improve. Activity patterns, far from being random, are closely related to temperature, humidity, wind, predators' activities and other factors. When conditions are favourable, literally hundreds of scorpions may be seen in prime habitats. With the first rains of the season, a great number of individuals may emerge simultaneously.

Although mostly nocturnal, a few species may also be active during the daytime. *Hottentotta conspersus*, for example, has been recorded wandering about in the shade of vegetation during the day, as have *Parabuthus stridulus* and *Opistophthalmus carinatus*. This, however, does not represent their usual pattern of activity. The only truly diurnal scorpion is *P. villosus*, which is commonly seen wandering about during the early morning and late afternoon, perhaps because of the availability of daytime prey.

> **In coastal areas in Namibia, *Parabuthus villosus* has been seen to drink condensed fog off grass stems. These areas – less than 110 km from the sea – may receive as little as 7–64 mm of rain annually, but as much as 161 mm of precipitation from fog.**

Conserving water

Scorpions use a number of physiological and behavioural mechanisms to conserve water. They achieve minimal water loss through their nocturnal tendencies, low levels of activity and shelter selection. Their bodies are designed to conserve water by means of lipids in the exoskeleton. Water is lost only via the spiracles of their booklungs and through their faeces. In times of environmental stress, a scorpion can close off its spiracles in order to conserve water.

Scorpions are able to drink freestanding water and often drink from pools at the beginning of the rainy season when they are thirsty. Some scorpions can live without drinking at all: the food they eat provides enough water for them to survive.

Regulating temperature

Scorpions are obliged to regulate their body temperature by external means. Unlike humans, they cannot lose heat by way of sweat glands and are unable, also, to generate heat. Instead, they avoid the hot sun and are active at night. When scorpions are cold, they are sluggish and vulnerable to predators. Research on various species in different habitats indicates that their activity levels peak at temperatures between 32°C and 38°C.

A perfect place for *Hadogenes* species to lie low and avoid the heat of the sun.

Shelter selection is of major importance to survival. Scorpions provide themselves with their own, more even, microclimate by exploiting shelters such as rock cracks, burrows and surface debris.

Discarded millipede rings are a sure sign that a scorpion lives under this rock.

INTERACTION

Scorpions are not normally social. They interact with other scorpions only in courtship and in rearing young (and, sometimes, in predation) and with other animals in predation. Formidable predators themselves, they might appear venomous enough to deter other animals, but this is hardly the case. Birds and even other scorpions are among the many that prey on scorpions.

Certain scorpion species tend to avoid other species by being active during specific times of night or only at times not favoured by the dominant species. In the Kalahari Desert, some scorpion species keep clear of the highly venomous and dominant *Parabuthus granulatus*. The dominant species may abound on one night. On other nights, other scorpions are found in the area.

The bond between the female and her young is also a fascinating aspect of scorpion behaviour. Under abnormal conditions, the young may remain with their mother even after moulting for the second time.

BURROWS

Deep burrows are often constructed by burrowing scorpions in very hot and dry areas, such as Namibia. As a rule, the deeper the burrow, the cooler and more humid the environment within the burrow. During the winter months, many scorpions shelter in their retreats and remain almost inactive. When conditions are unfavourable, the scorpion often seals off the burrow by pushing out substrate from inside to close the entrance. Sealed-off burrows are extremely difficult to locate.

Research has shown that when surface humidity is less than 5 per cent, the humidity of a burrow descending to a depth of 20 to 35 cm can be as high as 55 to 70 per cent. Burrow temperatures are also more constant than surface temperatures.

Burrow construction is a time-consuming and energy-sapping task and, once a burrow is constructed, it is not, under normal circumstances, abandoned. It has been estimated that a scorpion moves about 200 to 400 times its body weight in substrate when excavating a burrow. Field studies of North American species indicate that the majority of scorpions that shelter in burrows do not stray far from the entrance. Some move a considerable distance away, but invariably return to the burrow.

How the scorpion finds the burrow's entrance on its return is the subject of many theories. Scorpions are said to orient themselves by landmarks and other visual clues, wind direction, scent trails and even to navigate by the stars. At present there is no conclusive answer to the question – another facet of scorpion behaviour that we have yet to understand.

By burrowing into the sand to a depth where temperature and humidity remain fairly constant, *Opistophthalmus holmi*, a sand specialist, escapes the extremes of the Namib desert sand dune system, where surface temperatures can rise as high as 80°C at midday. Scorpions such as *O. capensis* escape the fires that occur frequently in summer in the fynbos regions of South Africa by retreating into their burrows. In the same areas, a surface-dwelling scorpion, *Uroplectes lineatus*, is at the mercy of the fire.

Fan-shaped substrate outside the burrow of *Opistophthalmus pugnax* (left, top) shows that its inhabitant has been spring cleaning. Note the enlarged part of the burrow (left) for resting or consuming prey. *Cheloctonus jonesii* has a distinctive deep burrow (right).

A hungry *Opistophthalmus holmi* in typical hunting pose.

Predation

Scorpions employ one of two techniques to catch prey. Species such as *Uroplectes otjimbinguensis*, *U. vittatus* and *Opistacanthus asper* seek out their prey. These three species are tree-living and forage actively at night. *Parabuthus granulatus* also forages for prey, travelling considerable distances in a single night. *P. granulatus* and *P. transvaalicus* are known to be attracted by movement, individuals often stopping to investigate moving objects before continuing on their way. Most scorpions, however, are sit-and-wait hunters and prefer to ambush prey that may walk or fly past them.

Sit–and–wait hunters

Sit–and–wait hunters (including many *Parabuthus* species) ambush prey from under vegetation or (like *Opistophthalmus* and *Cheloctonus* species) wait at the entrance to their burrows 'door keeping'. When a scorpion is hunting, it remains still with its pincers flexed open, resting on the substrate. As the prey wanders about, its movements create vibrations in the substrate and tiny air currents, which the scorpion 'feels' with microscopic slit sense organs on its legs and tail and the special hairs on its tarsi. It also has small sensory hairs in pits on the exoskeleton. By accurately monitoring the wave characteristics of vibrations, the scorpion is able to locate and capture prey on and below the surface.

A large adult *Hadogenes troglodytes* eating a centipede of the same size as itself. In different circumstances, the scorpion may well be the prey and the centipede the predator.

Vision does not play a large role in prey detection. Position of feet, on the other hand, is important. A scorpion's feet are arranged in a circular pattern, enabling different feet to sense vibrations at different times. As the prey moves closer, the scorpion rotates itself to get a better idea of which direction it is coming from and how far away it is. The human equivalent is hearing a sound and turning the head to determine the direction the sound is coming from.

As the prey comes within striking distance, the scorpion lunges towards it and grasps it with its pincers. Whether it uses its sting depends on the size of the prey and the strength of the scorpion's pincers. If the prey is large or struggles violently when held, the scorpion may inject venom into it to subdue it. Scorpions with weak pincers generally subdue their prey with venom; those with powerful pincers often crush their prey to death.

WHAT (AND HOW) SCORPIONS EAT

The opportunistic hunting strategies of scorpions result in a wide range of prey items, their varied diet consisting of just about any animal they can overpower. This includes insects, spiders and other scorpions. *Opistophthalmus carinatus* has been recorded eating earthworms. Some of the larger species will prey upon small vertebrates, gastropods, small reptiles, mammals and amphibians if the occasion arises.

Captured prey are eaten on the spot or taken to a burrow or chamber or a protected area under or on vegetation. Burrowing scorpions such as *Opistophthalmus pugnax, O. latimanus* and *O. glabrifrons* generally consume their prey inside their burrows. Those species in the genus that construct burrows under rocks and logs and other surface debris often make a small chamber near the entrance, which is where they usually consume their prey. *O. wahlbergii* is known to retreat to the terminal chamber of the burrow to feed. Before they feed, many scorpions orient the captured insect or other prey so as to consume it head first.

A gecko in the jaws of *Opistophthalmus capensis*.

Defensive behaviour

It seems logical that a venomous animal would use its venom as the primary defence against attack, in some cases physical attack by centipedes, other scorpions or predatory insects such as tiger beetles. However, many species of mildly venomous scorpions must defend themselves in other ways. *Cheloctonus* species, for example, live in burrows and will often interlock their pincers to shield themselves and block off the burrow to intruders. Many small species are well camouflaged to escape the notice of predators.

Studies suggest that scorpions adjust their activity patterns to protect themselves against predators by various strategies such as avoiding moonlit nights and keeping clear of scorpion species that prey on smaller species (sometimes even on smaller individuals of the same species). Many small species of scorpion spend much of their time in and around vegetation, where larger scorpions would have difficulty hunting. In some cases, they are active at times when larger or more aggressive species are inactive.

FOOD FOR OTHERS

In eco-systems such as deserts, where food is scarce, scorpions represent a stable food source for many animals because their populations remain constant for years, unlike insects such as locusts that are seasonal. They are preyed upon by a number of creatures, including hornbills, owls, frogs and even snakes, and by certain species of bat that specialize in ground-living prey. Many reptiles, such as monitor lizards and other large lizards, predate on scorpions, as do fellow arthropods such as large centipedes (Scolopendra), solifugids and spiders.

In a study of the diet of mammals in Botswana, the following animals were a few of those that were found to have eaten scorpions. In order of preference for scorpions as a food source, they are: honey badger, meercat, bat-eared fox, Selous' mongoose, Cape fox and yellow mongoose.

Other creatures employ some interesting techniques to subdue scorpionid prey. Solifuges, for instance, bite off the scorpion's tail before attacking the body. A foraging baboon will break off the tail of a scorpion before it eats the defenceless creature. Meercats employ the same technique. Large centipedes wrestle the scorpion by wrapping themselves around its body and biting it. Centipede venom is enough to subdue a scorpion effectively. Some animals are immune or resistant to scorpion venom, making scorpions easy to catch and kill.

Meercats rank high among animals that favour scorpions as food items.

SOUND TACTICS

Some scorpions are able to stridulate (produce sound) to scare off attackers. All *Parabuthus* species except *P. distridor* have areas of fine granulations, similar in texture to sandpaper, on the upper surface of the first two or three tail segments. They scrape the end of the sting over these areas to produce a 'chick-chick' sound.

Many *Opistophthalmus* species are able to produce a hissing sound with the bristles on the inner surface of their mouthparts (chelicerae), which are often visible when they are eating. They rub these bristles against special ridges on the underside of the carapace.

The position of a scorpion's eyes is an indication of its ability to stridulate. If the median eyes are further back on the carapace, there is more room under the carapace for stridulating. *Opistophthalmus* species with a deep V-shaped groove on the carapace produce a louder sound than those with a shallow V-shape. Smaller species are less able to produce a sound loud enough to scare off attackers and, thus, are less likely to stridulate.

A warning from a large Kalahari *Parabuthus granulatus.*

Male-female interaction

An important area of scorpion interaction is that between males and females during courtship and mating, which generally take place in the warmer months of the year. Courtship is usually initiated by the male, who goes in search of females at this time. Females wait to be found. As a result, males are more exposed than females to the risk of predation and it is little wonder that in some scorpion populations males are few and far between. During courtship, the male may manoeuvre the female over a distance of 25 m or more. After mating he departs or, occasionally, is eaten by the female. Rearing of the young is left to the female. Patterns of scorpion mating and reproductive behaviour are described in detail in the next chapter.

Females and their young

The only other form of scorpion interaction is that between females and their young. Young scorpions generally stay within the protective custody of their mothers until they moult for the first time. The female is fierce in the defence of her young and carries them on her back until they are able to fend for themselves (see pages 40–41).

SCORPION VENOM

If there is one thing everyone knows about scorpions it is that they are venomous – a reputation that has resulted in their being unjustly persecuted from the dawn of time.

A scorpion's coloration is not an indication of its venomosity, the colour of scorpions of the same species sometimes varying from place to place. A good example of this is *Parabuthus granulatus* which generally has a dark brown to black body with pale legs, but is orange in the Kalahari Desert. Size or markings are not an indication of venomosity, but the thickness of the tail in relation to the pincers is.

Scorpions have a pair of venom glands in the venom vesicle. Medically important scorpions are usually large species that are capable of injecting relatively large amounts of venom from their venom glands. The size of the venom vesicle is not an indication of venom strength, but of how much venom the scorpion can produce, store and inject. The scorpion is in complete control when injecting its venom and can vary the amount of venom it injects.

VENOMOSITY RULE OF THUMB: Scorpions with weak pincers have thick and powerful tails. These scorpions capture prey with their venom (*Parabuthus* species being a good example). Scorpions with thin tails and powerful pincers (*Hadogenes* species are typical of these) can crush large prey items, but their venom is extremely weak.

How scorpions use their venom

The venom of a single scorpion may include several different neurotoxins. Each is thought to perform a particular function by targeting specific nerve cells. Scorpion neurotoxins are particularly targeted against certain types of animals. Some scorpion species have even evolved neurotoxins specifically designed for deterring vertebrates.

Why is scorpion venom so complex? The explanation lies in the fact that scorpions use their venom for more than one purpose. The most obvious is prey capture. Whereas scorpions with powerful pincers may simply crush their prey, those with weaker pincers grasp their prey and sting it several times until it is subdued. By using its venom, a small scorpion is able to catch prey much larger than itself.

Venom is also used as a defence mechanism against animals such as honey badgers,

The sharp end of the tail with a drop of venom oozing out of the venom duct.

which prey heavily on scorpions. Some ingredients in scorpion venom do no more than induce pain. Although this is of little benefit when capturing prey, it is a huge advantage when warding off would-be predators. However, many mammals in the same evolutionary race are immune to scorpion venom and some, for example meerkats, suffer little from a scorpion sting.

A third use of venom is in mating. Males of certain *Hadogenes* species sting the female during courtship. When the female is stung by the male, she calms down and relaxes. It is not known whether venom is injected during this stinging action, but it is likely that some male scorpions possess pheromones that induce female receptivity.

Scorpions are immune to their own venom because of the toxin-resistant properties in their hemolymph. However, a scorpion will quickly die if venom is injected into its nerve ganglion (brain). Larger species are able to overcome smaller ones by the large quantity of venom they can inject in relation to the body size of the victim. Scorpion species that prey upon scorpions of the same size or larger are likely to have a neurotoxin in their venom specifically designed for fellow scorpions.

VENOM SCALE
Southern African Species
Indicates strength of venom, not toxicity (which depends also on quantity of venom injected).

WEAK	
1	Hadogenes
2	Cheloctonus, Opistacanthus
3	
4	Opistophthalmus (Bee sting)
5	(Wasp sting)
6	Uroplectes, Karasbergia, Lychas, Pseudolychas, Afroisometrus, Lisposoma
7	
8	Hottentotta
9	Parabuthus
10	Parabuthus
STRONG	

THE DANGER OF SCORPION VENOM

Drop for drop, scorpion venoms are some of the most powerful known, comparable to snake neurotoxins and exceeded in strength only by certain bacterial toxins. However, there is much variation in venom strength between species and the venom of different populations of the same species can show marked differences.

In southern Africa, the most venomous scorpions are contained in the genus *Parabuthus*. Three species stand out in medical importance – *P. granulatus* (our most venomous scorpion), *P. transvaalicus* and (less so) *P. mossambicensis*. Although responsible for few deaths, *P. transvaalicus* can spray venom. See pages 76–77 for information on the effect of scorpion venom on humans.

SCORPION BEHAVIOUR

MATING AND REPRODUCTION

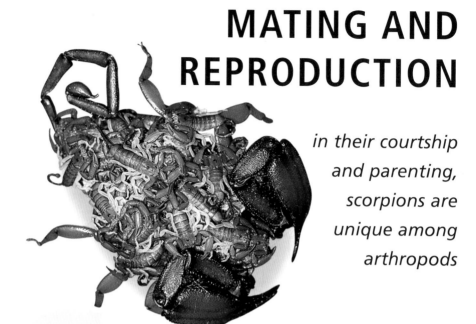

*in their courtship
and parenting,
scorpions are
unique among
arthropods*

SCORPIONS ARE UNIQUE in many ways. The scorpion 'dance' in which the male moves the female over a considerable distance is a well-known phenomenon. The young of scorpions are born alive – also unusual among arachnids, all of which, except for one family of mites, lay eggs.

In southern Africa, female scorpions give birth to their young during the warmer months of the year. The young generally remain with the mother until the first instar. Birth takes place in protected places such as burrows, crevices and under rocks and, consequently, is rarely observed in the wild.

MATING

Courtship

Courtship is usually initiated by the male. Males often wander about actively in the warmer months looking for females. Evidence suggests that the female gives off a pheromone that the male recognizes and responds to. Females remain in or near their shelters – a burrow, crevice or under a rock, depending on the species. During this time, males often inhabit temporary shelters not normally associated with the species.

When a male locates a female, he communicates his intentions through vibrations. He may judder his entire body, tap his pincers or wag his tail. These actions produce vibrations that travel through the substrate. *Opistacanthus asper* males tap their pincers on the trees where they live. Males of other species may club the substrate with their tails. This prenuptial communication from the male is important in informing the female that he is not a tasty meal. Only after the male is sure he has made his intentions known, and the female makes a positive response, will he approach her.

Securing the female

Males of most genera have modified pincers on their pedipalps, which are used to grasp the female. These modifications may be different for different genera. *Opistophthalmus* males have elongated pincers; those of *Parabuthus* males are bulbous. Some male *Uroplectes* have a tooth on the inside curvature of the pincers. *Cheloctonus* and *Opistacanthus* males have a hump on the inside margin of the pincers, which is used to grasp the female securely. (See page 73 for drawings of the pincers of southern African species.)

The male may grasp the female by her pincers or, in some cases, males and females lock mouthparts. Once the male has a firm grip on the female, he manoeuvres her to a place where he can deposit his spermatophore. He fans out his pectines to feel the substrate while looking for a smooth, hard place such as a rock, stone or branch to which to attach his spermatophore.

This part of courtship may last from five minutes to half an hour, during which time the male may drive the female as much as 25 m or more. One of them may use the end of its tail to club the other, with the sting folded away. In some instances, the male grasps the edge of the female's carapace with his mouthparts or grasps her at a joint in the pedipalps. He may also massage the female with his mouthparts.

Intent on mating, an *Opistophthalmus pugnax* male (left) and female (right).

Opistophthalmus pugnax male (left) and female (right) with mouthparts locked.

MATING AND REPRODUCTION

SEXUAL STING

In some *Hadogenes* species, the male may puncture the body wall of the female with his sting. It is not known whether venom is injected into the female, but the sexual sting appears to have a calming effect on her. As the female's pincers are powerful enough to crush the male quite easily, and as courtship often occurs in narrow cracks, the male must reach around the side of the female's body to sting her.

Not surprisingly, males of all except one *Hadogenes* species have incredibly long tails. The males's sting may remain in the female from 3 to 20 minutes or more, during which time the male carefully manoeuvres the female to a suitable place where he can deposit his spermatophore.

A necessary tool – the tail of a male *Hadogenes troglodytes*.

Positioning the female

Once the male has attached his spermatophore to a suitable object, he has to position the female so that the hooks on the top of the spermatophore catch the female's genital opening. After the male has deposited his spermatophore, he manoeuvres the female into position over it. The female may arch her body over the spermatophore or the male may lift her up and then drop her down. At this time, the female spreads the genital opercula, which normally cover her genital opening, and the end of the spermatophore enters her. The weight of the female bends the spermatophore and triggers the release of the sperm. She remains motionless on the spermatophore for just a few seconds or for a few minutes.

After sperm uptake, both male and female break away from each other, often violently. Either one may club the other with its tail, or probe the other with its sting. The female may try to eat the (smaller) male at this point. Afterwards, the spermatophore may be eaten by either male or female. Males are capable of mating more than once during the breeding season and can produce a new spermatophore in as little as six days. Females of many species are able to store sperm, and can produce more than one batch of young from a single mating. Courtship may take place even if the female is still carrying her young on her back.

Spermatophore emerges as an upside down T-shape with a sticky base at one end and a hook-like structure at the other. It is extruded from the male's genital opening in two halves. As the two halves emerge, sperm is added and the two halves are 'glued' together.

GESTATION AND BIRTH

The period of gestation varies according to species, and ranges from only a few months (in *Uroplectes*) to as long as 18 months (in *Hadogenes*). Such a long gestation period has no equal in the arachnid world. Developmental periods vary not only with species but with environmental factors such as temperature and the availability of prey. The young of scorpions are born live.

In difficult conditions, female scorpions are able to halt the development of their embryos or even to reabsorb them. Development resumes once conditions improve. In some species, substantial embryonic development does not occur during winter and other inactive periods. The sudden availability of prey in spring and summer may boost development, resulting in synchronized births. Synchronization increases the number of scorpions reaching sexual maturity since a larger number of individuals escape predation than would be the case if births were spread out over a longer period of time.

There are two types of embryonic development: viviparous (the young developing within the ovari-uterus of the female) and ovoviviparous (hatching from eggs within the body of the female). There are two corresponding methods (katoikogenic and apoikogenic) of embryonic development.

Katoikogenic scorpions do not produce eggs. Embryos develop in much the same way as human embryos with a placental attachment. Southern African katoikogenic scorpions include Scorpionidae (*Opistophthalmus*) and Ischnuridae (*Hadogenes, Cheloctonus, Opistacanthus*).

Apoikogenic scorpions develop from eggs, the developing embryos gaining their nourishment from the large amounts of yolk inside the egg. These scorpions are born within a birth sac. Southern African apoikogenic scorpions include Buthidae (*Uroplectes, Karasbergia, Lychas, Pseudolychas, Hottentotta, Parabuthus, Afroisometrus*) and Bothuridae (*Lisposoma*).

Giving birth

Females often seal off their retreats in preparation for birth. Shortly before giving birth, the female elevates the front part of her body well above the substrate. (This posture is known as stilting and also occurs when scorpions are uncomfortably warm.) The tail is

MATING AND REPRODUCTION

arched over the body and the pincers are flexed. The first two pairs of legs are held underneath the genital opening to form a birth basket for the emerging young. As the female stilts, the genital passage opens and the young emerge one by one. The birth process may take several hours or even one or more days.

THE YOUNG

Shortly after birth, young scorpions become active and climb onto the back of the female, who then resumes her normal posture. Katoikogenic scorpions are always born tail first, whereas apoikogenic scorpions are randomly orientated. In the case of apoikogenic scorpions, unfertilized eggs may also be deposited at this time.

Many factors influence litter size, including the size and nutritional state of the mother and, in some cases, the size of the young. There are usually about equal numbers of males and females in a litter. In *Parabuthus transvaalicus*, a litter of 32 consisted of 17 females and 15 males. In comparison with spiders, scorpions produce small litters. However, scorpions invest far more time and effort in maternal care, ensuring a greater survival rate.

Although newborn scorpions resemble adults, they are very pale and almost grub-like and are not fully developed. Young apoikogenic scorpions are bloated with yolk reserves. They do not so much feed as absorb stored nutrients from their bodies. Katoikogenic scorpions are characterized by proboscis-like mouthparts which are used by the female to transfer nutrients to the developing embryos in her body.

Below is a table of recorded litter sizes for southern African scorpions based on observation. Litter size may vary, however, with environmental factors.

SCORPION LITTERS
(average size given)

Parabuthus transvaalicus	32
Cheloctonus jonesii	28
Pseudolychas pegleri	28
Opistophthalmus pugnax	25
Uroplectes planimanus	22
Opistacanthus asper	22
Lychas burdoi	17
Hottentotta trilineatus	12
Uroplectes insignis	12
Uroplectes lineatus	8

Care of the young

The young stay within the protective custody of the female until they moult for the first time (the first instar). This period lasts from 9 to 14 days depending on the species and on environmental factors. To protect the young, whose soft exoskeletons lack the waxy lipids required for water conservation, females carrying young do not hunt or venture out of their temperature- and humidity-controlled retreats. In this early stage, the young maintain hydration by absorbing water from the mother's body.

The female is fierce in defence of her young, which are very vulnerable to predators because of their small size, conspicuous coloration and soft exoskeletons. The mother-child bond is maintained by scent. If a young scorpion falls off its mother's back, she will pick it up and help it regain its position. Excellent parental care is one reason why scorpions have met the challenge of survival so successfully.

After the first instar, the young scorpions begin to look more like miniature adults in coloration and proportions. The young now venture away from the female for the first time and may disperse into adjacent habitats. Although behaviour and shelter selection is fairly constant for adults of a species, juveniles often display deviant behaviour in, for instance, their choice of shelter.

The tube-like proboscis through which they feed is clearly visible on these newborn *Opistophthalmus karrooensis*.

First instar *Opistophthalmus glabrifrons* on their mother's back.

SOUTHERN AFRICAN SPECIES

southern Africa is blessed with a highly diverse and abundant scorpion fauna

MORE THAN 130 of the 1 500 scorpion species known in the world occur in Africa south of the Zambezi. Many species are endemic (occur nowhere else), several displaying specialized physical and behavioural characteristics. There have been many changes in the classification of southern African scorpions over the last few years and it is likely that there will be more.

Southern African scorpion species occur in four families:
- **BOTHURIDAE:** 1 genus
- **BUTHIDAE** (some of the world's most venomous scorpions): 7 genera
- **ISCHNURIDAE** (some of the world's least venomous scorpions): 3 genera
- **SCORPIONIDAE:** 1 genus

Scorpions occur in many different parts of the southern African region and in a variety of habitats. Some are adapted to life in trees, others live in burrows or rock crevices and yet others are generalists, taking advantage of a variety of shelters. In many cases, scorpion species show a preference for highly specific habitats.

The most species-rich scorpion locations in southern Africa are north-western, central and southern Namibia and the Namaqua highlands and the Richtersveld in South Africa. These areas often coincide with regions of rugged terrain such as mountain ranges.

Other scorpion hotspots are the:

- Mpumalanga escarpment
- Eastern Highlands (Zimbabwe)
- Southern Kalahari
- Southern Cape fold mountains
- Magaliesberg, North-West Province
- Waterberg-Soutpansberg complex
- Limpopo depression

KwaZulu-Natal dune forests (above) host large scorpion populations. In the west, the arid Richtersveld (below, right) contains many species. Scorpions occur in dry, sandy, semi-vegetated areas (bottom left) and in surface debris in bushveld (below, left). Very specialized species are to be found in the Namib sand system (bottom right). Scorpions do not occur on high Lesotho peaks (right).

There are many reasons for such widespread distribution and diverse habitat selection other than the remarkable ability of scorpions to adapt to an environment. Their limited means of dispersal, meanwhile, has resulted in a high degree of speciation and specialization.

In unfavourable conditions, scorpions may be few and far between. In good times, however, a locality may teem with scorpions and it is not uncommon to see more than a hundred individuals on a single night.

OVERVIEW OF SOUTHERN AFRICAN SCORPIONS

Genus	Species	Distribution	Habitat	Venomosity
BOTHURIDAE				
Lisposoma (endemic; 2 species)	L. elegans (up to 28 mm)	Northern and central Namibia.	On open ground and under small rocks on hard surfaces.	Painful sting, not medically important.
	L. josehermana (rare) (up to 32 mm)	Otavi Highlands, Namibia.	Deep burrows, under large boulders in damp soil, in vegetated areas and caves.	Painful sting, not medically important.
BUTHIDAE				
Afroisometrus (1 species)	A. minshullae (rare) (up to 30 mm)	Described from Zimbabwe.	Granite outcrops.	Painful sting, not medically important.
Hottentotta (3 species)	H. arenaceus (up to 43 mm)	Southern Namibia into Northern Cape.	Scrapes or burrows under dense vegetation.	Very painful sting.
	H. conspersus (up to 62 mm)	Northern Namibia into Angola.	Under large rocks or boulders in sandy areas, under logs or bark of dead trees on the ground.	Very painful sting.
	H. trilineatus (up to 60 mm)	North of Soutpansberg into Zimbabwe, Mozambique and Zambia.	Under rocks and logs in sandy areas.	Very venomous, medically important.
Karasbergia (endemic; 1 species)	K. muthueni (up to 23 mm)	North-western Cape, southern Namibia.	Hard gritty soils.	Painful sting, not medically important.
Lychas (1 species)	L. burdoi (up to 40 mm)	North-eastern Kruger, eastern Zimbabwe, Mozambique into central Africa.	Leaf litter or vegetation in moist habitats.	Venom potent but normally not life-threatening.
Parabuthus (20 species)	(from 70–180 mm; P. leavipes up to 50 mm)	Widespread in the subregion, mainly in arid and semi-arid areas.	Burrow in open ground and also under stones or rocks, many in restricted habitats.	Venom life-threatening; most venomous scorpions in southern Africa.
Pseudolychas (endemic; 3 species)	P. pegleri (up to 45 mm)	Eastern South Africa and Zimbabwe; Mozambique.	Moist habitats, under rocks and surface debris; known to enter houses.	Painful sting, not medically important.
Uroplectes (19 species)	(average length 50 mm)	Throughout southern Africa; the most widespread genus in the region.	In trees, under stones and logs, in sand at base of bushes and grass tufts.	Painful sting, not medically important.

Genus	Species	Distribution	Habitat	Venomosity
ISCHNURIDAE				
Opistacanthus (6 species; including *O. capensis, O. asper, O. validus*)	*O. capensis* (up to 90 mm)	South-eastern Cape.	Under rocks and other surface debris.	Very docile, venom weak.
	O. asper (up to 100 mm)	Northern KwaZulu-Natal and Mpumalanga into Zimbabwe and Mozambique.	Under loose bark and fissures in trees.	Very docile, venom weak.
	O. validus (up to 90 mm)	Lesotho and eastern South Africa, including the Drakensberg.	Under rocks and other surface debris.	Very docile, venom weak.
Cheloctonus (5 species; including *C. jonesii*)	(up to 90 mm)	Eastern part of the region.	Burrow in open ground, *C. jonesii* in peaty soils.	Very docile, venom weak.
Hadogenes (8 species)	(up to 210 mm)	Namibia to northern KwaZulu-Natal and north into Tanzania.	Mountain ranges and rocky outcrops, many in narrowly restricted habitats.	Seldom sting, venom very weak.
SCORPIONIDAE				
Opistophthalmus (endemic; over 59 species)	(up to 180 mm)	Throughout the region, especially Northern Cape and Namibia.	Construct burrows in various habitats.	Painful sting, venom normally not life-threatening.

- In southern Africa, there is a distinct divide between the eastern and western scorpion fauna.
- Some southern African *Opistophthalmus* species have a total distribution of less than 20 km^2.

FAMILY BOTHURIDAE

Genus *Lisposoma*

Contains only two species, both endemic to northern and central Namibia. Both species have thick tails and robust pedipalps. They are rarely seen because of their small size and limited distribution. No envenomations are recorded. Because of their size and the small amount of venom they produce, their sting is unlikely to be of medical importance.

Lisposoma elegans

Up to 28 mm long. Males are smaller and more slender than females and have more bulbous pedipalps. Has been collected (using a UV light) from gritty, open ground, where individuals are to be found resting. Also under small rocks on hard surfaces. It is not known whether this species constructs burrows.

Lisposoma josehermana

Up to 32 mm long. It is found in deep burrows under large boulders half imbedded in slightly damp ground, on south-facing slopes. Also in fairly dense mixed mopane and acacia woodland in vegetated areas and in shallow caves. Its area of distribution is the Otavi Highlands in the Tsumeb and Grootfontein districts (Namibia). It is difficult to find during the day.

FAMILY BUTHIDAE

Genus *Afroisometrus*

A recently created genus which contains only one species, described from Zimbabwe in 1994 and previously placed in the genus *Lychas*. This scorpion was named in honour of the late Jacqueline Minshull who founded the department of arachnology at the Bulawayo Museum in Zimbabwe. Because of its small size and distribution, it is unlikely to be encountered.

Afroisometrus minshullae

Up to 30 mm long. Yellowish, lacks a subaculear tubercle. Males are smaller than females. One of the few southern African scorpions with six eyes (two median eyes and two pairs of large lateral eyes on either side of the carapace). Because of its coloration and small size, this species is best located with a UV light. It seeks shelter on the southern side of boulders on granite outcrops. This scorpion may also be found under rocks and logs in such areas. The small number of collection records indicates that it is a rare species.

Genus *Hottentotta*

There are three southern African species in this genus (previously named *Buthotus*). They are medium-sized and robust, with stout appendages. Males are slightly smaller and more slender than females, and have more bulbous pincers. All species are characterized by lyre-shaped markings on the carapace and three distinct longitudinal keels on the tergites. They do not stridulate in any way. This genus is considered medically important as all its members can inflict a very painful sting (although not as venomous as that of some *Parabuthus*). They occur in distinctly different areas: two species in Namibia, the Northern Cape and Angola; the third in South Africa north of the Soutpansberg and Zimbabwe, Mozambique and Zambia.

Male *Hottentotta* (right) are smaller than females (left) and have more bulbous pedipalps.

Hottentotta trilineatus

Up to 60 mm long. Orange-brown and well camouflaged in the hot, sandy areas it inhabits. The name refers to the three (tri) lines on the tergites. It constructs a scrape under rocks, logs and other surface debris. Potent venom and a fiery temperament make this species medically important.

H. trilineatus, Kruger Park.

Hottentotta arenaceus

Up to 43 mm long. Pale orange-yellow, southern-most populations are slightly darker. Often constructs burrows 6–10 cm deep at the base of bushes in sandy environments. Males are known to wander at night, sheltering under bushes during the day. Females generally stay on or near vegetation. Because of its small size and cryptic coloration, it is best located with a UV light at night. May be seen at night resting on raised sandy surfaces and lying in wait for small insects on vegetation or foraging on bushes. Very common in certain areas.

H. arenaceus, Fish River Canyon.

Hottentotta conspersus

Up to 62 mm long. Dark yellow to brown. In the northern parts of their range, many females have almost smooth and shiny sternites. Found sheltering under large rocks or boulders (its pre-ferred habitat) in sandy areas or under logs or the bark of dead trees on the ground. Individuals have been observed wandering about in the shade of trees during the day.

Genus *Karasbergia*

Contains a single species, *K. muthueni,* which is endemic to southern Africa and is found in the north-western Cape and southern Namibia in areas of hard gritty soils. Very little is known of the life history of this scorpion, which is unlikely to be encountered because of its distribution and very small size. Venom is probably potent but in small amounts and, hence, unlikely to be life-threatening. There are no recorded envenomations. Because of its size, *K. muthueni* is easily mistaken for juvenile *Parabuthus.* Its behaviour, however, is more casual and like that of mature *Parabuthus.* (Juvenile *Parabuthus* are very active and defend themselves vigor-ously, stinging, running away and stridulating.)

Karasbergia muthueni

Up to 23 mm long. Males have a longer and more slender tail than females, more slender pedipalps and a more granular carapace. Males have 14–16 pectinal teeth, females 11–13. One of the few southern African scorpions with six eyes (two median eyes and two groups of two lateral eyes). Occasionally with eight eyes (two median and two groups of three lateral eyes), but one group of lateral eyes is often underdeveloped and reduced in size. Diagnostic features include very round, almost bead-like caudal (tail) segments. Its small size and cryptic coloration make it virtually impossible to find during the day. The only practical method of location is with a UV light at night (which results in the capture of more males than females). *K. muthueni* does not stridulate, has never been collected from sandy substrates and has not been observed to construct burrows or scrapes.

Genus *Lychas*

Only one species, *L. burdoi*, has been collected in southern Africa. (About 50 species occur in southern and central Africa, South America, Australia and Asia.) Venom is strong, but not life-threatening under normal circumstances. *L. burdoi* is often mistaken for a *Uroplectes* species, but has a very large subaculeur tubercle. *Uroplectes* are indeed similar, but usually have a small subaculeur tubercule.

Lychas burdoi

Up to 40 mm long. Characterized by yellow and black markings, its coloration providing the perfect camouflage in moist leaf litter or loose bark. Males of the genus are smaller than females. This species is found in moist habitats in the far north-eastern corner of the Kruger National Park, eastern Zimbabwe and Mozambique, and up into central Africa. Because of their small size and cryptic coloration, *Lychas* species are best collected at night with the aid of a UV light.

L. burdoi, Kruger Park.

Genus *Parabuthus*

There are 20 *Parabuthus* species distributed throughout southern Africa, predominantly in arid and semi-arid regions. These large scorpions (mostly between 70 mm and 180 mm in length) are the most venomous in southern Africa. Their distribution is influenced by rainfall, and they generally occur in areas that receive less than 600 mm of rain annually. Most species inhabit sandy regions.

Many *Parabuthus* species show colour variations across their distribution ranges. In some species, e.g. *P. leavipes*, the last two tail segments are darker. All, except *P. distridor*, have rough areas on the upper surface of the first segment of the extremely thick, strong and keeled tail. In many species, the second tail segment is also characterized by a rough area on the upper surface. By scraping the sting on these rough areas, a 'chick-chick' sound is produced. Pincers are smooth and weak. Many species have large venom glands. No *Parabuthus* has a subaculeur tubercle. Males are smaller and more slender than females, and use their more bulbous pedipalps to grasp the female's pedipalps during courtship.

All species are adapted for burrowing. Most construct burrows in the open ground, but *P. muelleri*, *P. planicauda*, *P. transvaalicus*, *P. villosus* and some others construct burrows under stones or rocks. The ridges on the fifth tail segment are used to loosen the substrate before dragging it away with the first two pairs of legs. Unlike *Opistophthalmus* (which carry the substrate outside as they burrow), *Parabuthus* shovel substrate behind them, making it look as though the burrow has been filled in. Gravid females often excavate deep burrows in which to give birth.

Many species show a narrow habitat tolerance. Several are adapted to living in the sand dune systems of the Namib and Kalahari deserts and have specialized body features for locomotion and burrowing in soft sand. The distribution of sand dune scorpions such as *P. kalaharicus* (found in shallow burrows at the base of vegetation or under rocks in sandy and gritty areas between the Kalahari sand dunes) is restricted as they are unable to burrow into hard substrate. *P. planicauda* and *P. villosus* are specially adapted for rocky habitats. This is evident in their flattened appearance and elongated appendages.

Parabuthus are often the dominant species where they occur, other scorpions avoiding the more aggressive species by being active at different times or by restricting their activities to vegetated areas. Many *Parabuthus* species ambush prey from under vegetation, but several are active hunters. One such species is *P. granulatus*, southern Africa's most venomous scorpion. *P. villosus* (the largest Buthidae in the world) is unique among scorpions in that it is active in the day.

As is typical among scorpions, the female *Parabuthus* (left) is larger than the male (right). *P. transvaalicus* shown here, Limpopo Valley.

Parabuthus villosus
Up to 180 mm long, the largest member of the family in the world. A very hairy scorpion, pitch black to brown with yellow legs depending on the location within its distribution. It is active during the day and is often seen wandering about in the early morning and late afternoon. Found in rocky areas, in scrapes under surface debris such as rocks and logs and, on occasion, under the loose bark of fallen trees.

P. villosus, Augrabies Falls.

Parabuthus brevimanus
Up to 50 mm long, one of the smallest in the genus. Digs shallow burrows at the base of shrubs in sandy areas, and is occasionally found under rocks in a shallow scrape. This species is widespread throughout Namibia and adjacent areas in the Northern Cape. Although its venom is likely to be potent, it is not of medical importance because of its size and the small amount of venom it injects. No envenomations have been reported.

P. brevimanus, Naukluft Mountain Zebra Park.

Parabuthus granulatus
Up to 160 mm long. Coloration is variable, ranging from overall orange (Kalahari Desert) to blackish brown (Postmasburg, Northern Cape) to brown with pale legs (Middelburg, Eastern Cape). Characterized by a small sting. Digs burrows in consolidated sandy soils at the base of shrubs. A very aggressive predator that will prey upon other scorpion species; one of the few scorpions that forages actively for prey. Is known to enter houses or seek shelter adjacent to dwellings. This is southern Africa's most venomous scorpion.

P. granulatus, Kalahari Desert.

Parabuthus mossambicensis
Up to 80 mm long. Because of its reddish brown to orange coloration, it blends into the reddish sand where it lives. Burrows to depths of 10–30 cm under surface debris or in the open. Its distribution and the strength and quantity of its venom make it medically important. Has been known to spray venom.

P. mossambicensis, **northern Kruger Park.**

Parabuthus leavipes
Up to 50 mm long, one of the smallest in the genus. Its last two tail segments are blackened. Digs shallow scrapes under rocks in consolidated sand and gritty soils. The distribution of this species intrudes marginally into sand systems. It is not common. Venom is potent, but no human envenomations have as yet been reported.

P. leavipes, **Keetmanshoop, Namibia.**

Parabuthus raudus
Up to 160 mm long. Similar in appearance to *P. schlechteri*. Digs a shallow burrow at the base of vegetation in sandy soil, and is occasionally found on rocks, logs and other surface debris and on shrubs. Has never been found to shelter under loose bark of fallen trees or logs. This is the largest and most common scorpion in the Kalahari sand system. It has not been recorded in the Namib sand system. This species has been known to spray venom when extremely provoked.

P. raudus, **Kalahari Desert.**

Parabuthus schlechteri

Up to 140 mm long. Similar in appearance to *P. raudus*, but has 10 granular keels on the fourth caudal segment. Digs shallow scrapes under rocks in areas of consolidated sand and gritty soils. Although widely distributed, it is not often encountered. This species has been known to spray venom when extremely provoked.

P. schlechteri, Karasberg, southern Namibia.

Parabuthus capensis

Up to 105 mm long. The colour form shown here is typical, but there is also a black form found near the coastal areas north of Cape Town. Usually found in scrapes under rocks, but may shelter under almost any kind of surface debris. Has been recorded inside dwellings. It occurs in populated areas of the Western Cape and has been implicated in a number of envenomations, but is not known to have caused any human deaths.

P. capensis, Worcester, Western Cape.

Parabuthus namibensis

Up to 100 mm long. The two last tail segments and sting are blackened. Because of its remote and restricted distribution (in the central and northern coastal areas of Namibia) very little is known about its habits and preferred habitat. It has been found on gravel plains in sandy, gritty areas, where it constructs burrows.

P. namibensis, north of Swakopmund, Namibia.

Parabuthus transvaalicus

Up to 150 mm long. Large and black. Venom vesicle is wider than the last caudal segment. Found in rocky habitats under surface debris, including logs, rocks and man-made 'shelters' such as rubble and garden refuse. Venom is potent and injected in large amounts, hence of medical importance. It sprays venom only when irritated.

Gravid *P. transvaalicus* female, Limpopo Valley.

Parabuthus stridulus

Up to 110 mm long. Easily identified by its shiny exoskeleton. It is one of the more common species in the Namib sand system and is known to wander about during the daytime. Digs shallow burrows at the base of small, sparsely vegetated sand dunes.

P. stridulus, Lüderitz, southern Namibia.

This species is distributed from the Ugab River in the north of Namibia to the Orange River in the south, and is often found in the intertidal zone within its range.

Genus *Pseudolychas*

An endemic southern African genus with three species, occurring on the moister eastern side of Zimbabwe, the eastern side of South Africa and in Mozambique. These species are found in vegetated areas under surface debris, where humidity is higher. The only species likely to be encountered is *P. pegleri*.

Pseudolychas pegleri
Up to 45 mm long. Reddish brown, often mistaken for a *Uroplectes* species because of its size, but can be distinguished by three ridges on the first four tergites. The shape of the sting is distinctive, the venom vesicle elongated and flattish with a short, recurved aculeus (like that of *Opistacanthus* and *Hadogenes* species). Does not construct a burrow, but makes a simple scrape under surface debris such as leaf litter, rocks and loose bark. Known to occur in highly urbanized areas, even industrial areas in Gauteng. Individuals have been found in houses – in sinks, bathtubs and showers. Sting is very painful, but not of medical importance.

P. pegleri, Randfontein, Gauteng.

Genus *Uroplectes*

Contains 50 species, distributed throughout southern and eastern Africa. There are 19 species known in southern Africa. Small to medium-sized (on average 50 mm long) with a fiery disposition and a painful sting. The smallest species is only 24 mm long, the largest reaches 70 mm. *Uroplectes* have the widest distribution of all scorpions in southern Africa and occur in habitats ranging from tropical dune forests (northern Natal) to deserts (Namibia). They are found in trees, under stones and logs and in sand at the base of bushes and grass tufts.

These brightly coloured scorpions range from yellow to brown to red or green, with areas of pigmentation, bands of coloration and other markings. Many *Uroplectes* species have a distinct subaculeur tubercle and the venom vesicle of the sting (telson) is distinctively shaped. Distinguishing males from females depends greatly on the species: in some species the first pectinal tooth is modified; in others the shape of the pedipalps is important. Males in this genus are smaller and more slender than females.

Uroplectes species run very fast, usually with their tails straight out. They are quick to defend themselves and their victims are usually stung repeatedly. They are not known to stridulate. Their venom is potent but, under normal circumstances, it is not life-threatening.

Uroplectes planimanus

Up to 70 mm long, the largest in the genus. It is characterized by wide pincers. Females have a sickle-shaped basal pectinal tooth. This species lives in rocky habitats and is often very common in prime habitats under and on top of rocks or under rock exfoliations.

U. planimanus, Limpopo Valley.

Uroplectes olivaceus

Up to 60 mm long. Characterized by a subaculear tubercle. The tail has fine granulations. Females are more granular and have a very long and broad basal pectinal tooth. It is not choosy about where it shelters and may be found under almost any available surface debris and under loose bark or in rock crevices. This species is responsible for a high number of envenomations and is one of the few found in shoes and clothing.

U. olivaceus, central Kruger Park.

Uroplectes carinatus

Up to 50 mm long. Dark yellow-orange to orange-brown, with a single dark line down the centre of the tergites. The basal pectinal tooth is broad in females. It is found in scrapes under rocks and surface debris in areas of hard substrate. This species is very widely distributed, but is absent from high rainfall areas in South Africa.

U. carinatus, near Clanwilliam, Western Cape.

Uroplectes vittatus

Up to 55 mm long. Similar in appearance to *U. triangulifer* but has smooth, more elongated tail segments. The last two tail segments are dark. Widespread, and is found under the bark of trees or under fallen logs. Often the culprit when people carrying wood are stung.

U. vittatus, near Gaborone, Botswana.

Uroplectes triangulifer

Up to 52 mm long. Similar in appearance to *U. vittatus* but its tail segments are not smooth. Each tergite has two broad, dark bands. Very widely distributed in grassland areas, where it makes a shallow scrape under rocks. This species has been known to enter houses in Gauteng, South Africa.

U. triangulifer, Klipriviersberg Nature Reserve, Gauteng.

Uroplectes formosus

Up to 24 mm long, the smallest species in the genus. Its cryptic coloration camouflages it perfectly against the bark of trees. Males have a raised bump on the inside margin of the pincers. Found under loose bark or in cracks and crevices in trees and fallen logs. Common in the dune forests and coastal areas of Natal. Like the other members of the genus, this species is eager to sting, but is not medically important.

U. formosus, Cape Vidal, northern KwaZulu-Natal.

SOUTHERN AFRICAN SPECIES

57

Uroplectes lineatus

Up to 45 mm long. This cryptically coloured scorpion occurs in coastal areas and is one of the more common species in those parts. It occurs in mountainous forest and fynbos vegetation types, where it can be found under rocks or loose bark and in rock crevices.

U. lineatus, Hermanus, Western Cape.

Uroplectes variegates

Up to 40 mm long. Cryptically coloured, distinguishable by seven longitudinal dark lines on the tergites. Constructs a clearly defined run about 13 cm long, ending in a small chamber. It is found in and around the Cape Peninsula up to Langebaan on the west coast.

U. variegates, Cape of Good Hope Nature Reserve, Western Cape.

Uroplectes insignis

Up to 45 mm long. It has a highly restricted distribution, limited to parts of Newlands Forest and Kirstenbosch in the Cape Peninsula. It is found under surface debris and has been reported to congregate in groups under rocks in winter. This may indicate a lack of over-wintering shelters rather than a form of sociality.

U. insignis, Newlands Forest, Cape Peninsula.

FAMILY ISCHNURIDAE

Genus *Opistacanthus*

Contains six species in southern Africa, in the eastern parts of the region. All are very docile. Males have a bump on the inside margin of the pedipalps. None stridulates. Members of this genus are similar in appearance to *Cheloctonus* species.

Opistacanthus asper

Up to 100 mm long. Uniformly black in the northern part of its range. Forages for food on trees and bushes at night, preferring dead acacias, in particular knobthorn (*Acacia nigrescens*). The texture of the bark of this tree provides crevices from where it can ambush prey that walks past. Males are known to tap their pincers to inform females of their intention to mate.

O. asper, northern Natal.

Opistacanthus validus

Up to 90 mm long. Blackish-brown and often muddy-looking grassland species with a wide distribution. It is found under rocks and other surface debris, where it makes a small scrape. It has been recorded at a high altitude (2 400 m above sea level) in South Africa, and is one of the few scorpion species found in Lesotho. Very docile, rarely using its sting in self-defence.

O. validus, Golden Gate National Park.

Opistacanthus capensis
Up to 90 mm long. It is found under rocks, where it may construct a scrape. Also under loose bark and other surface debris, and bricks, wood piles, doormats and other human debris. Often encountered at night, and has been known to enter houses. Very common in the coastal parts of its range, the most common scorpion species in Knysna (South Africa) and similar areas. Very docile indeed, rarely using its venom in self-defence.

O. capensis, Knysna forests.

Genus *Cheloctonus*

Contains five species of medium-sized scorpions, distributed throughout the eastern parts of the region. Males are more slender than females and have a prominent tooth on the moveable pedipalp finger, which is absent or reduced in females. Generally very docile and not of medical importance. Under normal conditions their venom has little or no effect except for mild pain and slight swelling. They often shield themselves when threatened by interlocking their pincers, effectively sealing off their burrows. They are closely related to *Opistacanthus* and look very similar.

Cheloctonus jonesii
Up to 90 mm long. Coloration black overall in northern Natal; brown with yellow legs in Mpumalanga. It is found in unmistakable near-vertical burrows in peaty soils, and avoids areas that become waterlogged. Burrows are situated in the open, the entrance often showing no sign of recently excavated substrate. A small chamber is constructed at the end of the burrow. In Jozini in South Africa, this species may reach population densities as high as four individuals per square metre.

C. jonesii, Sodwana Bay, KwaZulu-Natal.

Genus *Hadogenes*

Commonly called rock scorpions, with more than 17 large to very large species worldwide. They occur, in Africa, from South Africa north to Tanzania. This genus contains the longest scorpions in the world, notably *H. troglodytes*. They are easily identified by their very large pedipalps, thin tail, small venom vesicle, very flat appearance and elongated appendages. Their range is restricted to mountain ranges or discrete rocky outcrops. They are widespread across southern Africa, from KwaZulu-Natal in the east to Namibia in the west.

Well adapted to life in rocky environments, *Hadogenes* species even have compressed legs and tails to allow them to creep into narrow cracks and fissures in rocks. The tarsus is highly specialized for locomotion on hard substrate, providing a vice-like grip that enables the scorpion to walk upside down on rocks and to cling tenaciously to smooth surfaces. Both sexes possess large, powerful pincers and the female could easily crush the male during courtship. Males are easily recognized by their elongated tails, which are used in courtship. The last sternite is wider in females. *Hadogenes* males are often few in number and are hard to find. Males and females take about six years to reach sexual maturity.

Male (left) and female (right) *Hadogenes* of equal size. The male's tail is extremely elongated.

Hadogenes have such narrow habitat requirements that different species may be strictly separated by unsuitable conditions. A good example of this is in the Limpopo Province where *H. gunningi* forms a population in the Magaliesberg which is quite distinct from an *H. gracilis* population on rocky outcrops just 10 km to the north. Another good example is *H. minor*, which, in South Africa, is found only in the Cederberg and adjacent rocky outcrops. Studies show that females inhabit the same crack for many months, even years.

Shelters can often be identified by the remains of prey (millipedes, insects, smaller scorpions, spiders) outside the entrance. Being so large, *Hadogenes* are also able to prey on small vertebrates. Their venom is so weak that they seldom use their sting for defence or prey-capture. All species are strictly nocturnal, and lie in wait for prey at the entrance to the shelter.

Identifying *Hadogenes* species can be very difficult. As they are restricted, however, to distinct mountain ranges and different species seldom inhabit the same area, it is possible to narrow down possible species to just a few by using known distributions. Certain features, such as the shape of the front margin of the carapace and (in a few species) the tail, can be used to key out this genus.

Hadogenes are under great pressure from habitat destruction. Small distributions and long life cycles and gestation periods make them highly vulnerable. Lifespan in the wild is estimated at 25 to 30 years.

Hadogenes gunningi

Up to 70 mm long. Found throughout the Magaliesberg range and rocky outcrops in Gauteng. Inhabits cracks and rock exfoliations, its limited distribution encroaching on highly urbanized areas. Because of its distribution, this scorpion is in special need of protection from urbanization and habitat destruction.

H. gunningi female with second instar, Krugersdorp.

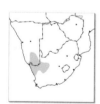

Hadogenes phyllodes

Up to 180 mm long. Males have extremely long tails. Relatively widespread for a member of the genus, but confined to southern Namibia and the north-western Cape. Its most northern distribution is the Bruckaros crater, Namibia. It inhabits cracks and crevices in rocks and rock exfoliations.

H. phyllodes, Augrabies Falls.

Hadogenes gracilis
Up to 130 mm long. It is found under and on top of rocks, in cracks and crevices. The distribution of this species is highly restricted in South Africa to scattered rocky outcrops in the North-West Province about 10 km north of the Magaliesberg range. Mining in the area has accelerated destruction of its habitat.

H. gracilis, Magaliesberg, North-West Province.

Hadogenes tityrus
Up to 80 mm long. Distributed throughout the southern third of Namibia in rocky and mountainous regions. Does not show the typical sexual dimorphism of the genus, the tail of the male not being longer than that of the female. Pincers in both sexes are elongated.

H. tityrus, Richtersveld, Northern Cape.

Hadogenes zuluanus
Up to 160 mm long. Coloration ranges from brown to dark brown. This species is found in mountain ranges and rocky outcrops in northern Natal and adjacent areas in Swaziland and Mozambique. Like other members of the genus, it shelters in crevices and cracks in rocks and rock exfoliations.

H. zuluanus, Jozini, northern KwaZulu-Natal.

SOUTHERN AFRICAN SPECIES

Hadogenes troglodytes

Up to 210 mm (males), the longest scorpion in the world. Has many colour variations, from black in southern populations (Pilanesberg National Park, South Africa) to brown with yellow legs in southern Mozambique. Distribution is restricted to mountain ranges and rocky outcrops, where it is found in cracks and narrow spaces. Small brood size (about 20) and a slow rate of reproduction (gestation period of up to 18 months) make this species particularly sensitive to destruction of its habitat. It takes 8–10 years to reach maturity.

H. troglodytes, Soutpansberg, Limpopo Province.

Hadogenes zumpti

Up to 150 mm long. Distribution is restricted to the northern Richtersveld. It has not been recorded north of the Orange River. Like others in the genus, it inhabits cracks in rocks and rock exfoliations. Its highly localized distribution makes it extremely vulnerable to habitat destruction.

H. zumpti, Richtersveld, Northern Cape.

Hadogenes minor

Up to 130 mm long. Its distribution is restricted to the Cederberg mountains in the Western Cape. Specimens from the northern parts of the range have yellow venom vesicles. Like other members of the genus, this species is rock-dwelling and inhabits cracks in rocks and rock exfoliations.

H. minor, Cederberg, Western Cape.

FAMILY SCORPIONIDAE

Genus *Opistophthalmus*

Endemic to southern Africa, with more than 59 recognized species, commonly known as burrowing scorpions. They are distributed throughout the region and are especially diverse in the Northern Cape and Namibia. This is an intriguing genus because of the high degree of specialization of some of its members. It contains some of the world's most beautifully coloured scorpions, medium to large (180 mm), with large robust pincers and a relatively thin tail. The sting is long and slender. Many have an obvious keel on the pincers.

In species such as *O. pallipes* and *O. longicauda* sexual dimorphism is extreme, with males displaying exquisitely elongated pedipalps. The pedipalps of females are generally more rounded. Males are smaller than females. Females construct deeper burrows.

Opistophthalmus use their mouthparts to excavate burrows and the first two pairs of legs to drag away the loose substrate. Many species burrow under surface debris, including rocks. Burrow configuration is not an indication of species, but is influenced by soil hardness, soil texture, submerged obstructions such as rocks and tree roots, degree of slope and other factors. The burrow usually spirals to the left as it descends. Active burrows are clearly visible because of the newly excavated substrate deposited in a fan shape and the scattered prey remains at the entrance. No member of the genus lives in trees.

Many *Opistophthalmus* species have narrow habitat requirements (influenced by the texture and hardness of the substrate) and localized distributions. A few species in the genus, however, are widespread, for example *O. wahlbergii*. *Opistophthalmus*

Male (bottom) *Opistophthalmus* are smaller and have more elongated pincers than females (top). *O. pallipes* shown here.

are sit-and-wait predators and often spend many daylight hours at the entrance to their burrows or 2–3 cm within. This door keeping is probably related to prey capture as daytime insects form a large part of their diet. They run out and grab at prey that wanders past, and will often grab viciously at an object such as a stick that is moved about at the burrow's entrance. Because their venom is mild, they tend to crush prey with their pincers, but will use their venom to subdue larger prey.

Many species stridulate using their mouthparts. They produce a hissing sound by rubbing the stiff bristles on the mouthparts against the underside of the carapace. Species with a deep V-shaped groove on the carapace produce a louder sound than those with a shallow V, and in some species, e.g. *O. pallipes*, even second instar individuals can stridulate loudly. Smaller species are less likely to stridulate.

Opistophthalmus can deliver a painful sting, but envenomation is not life-threatening in normal circumstances.

Opistophthalmus glabrifrons

Up to 120 mm long. Distributed over a wide range, the largest specimens found in areas such as Musina. Much smaller in areas north of the Magaliesberg in the North-West Province and Gauteng. Found in loamy soils, burrowing in the open or under surface debris to depths of 18–25 cm.

O. glabrifrons, near Musina, Limpopo Province.

Opistophthalmus pugnax

Up to 70 mm long. Muddy looking, characterized by corrugations on the last sternite. Males have corrugations on the last two sternites. Burrows are constructed under rocks and other surface debris. A very common species on rocky outcrops and ridges in Gauteng. It is not particularly aggressive and very rarely enters houses.

O. pugnax, Gauteng.

Opistophthalmus carinatus

Up to 110 mm long. Large and very robust, with powerful pincers. Burrow entrances are usually at the side of large stones and boulders, but may also be on open ground. In the Kalahari Desert, individuals can be found under large calcrete stones or dead vegetation. They may also be found under the loose bark of fallen trees.

O. carinatus, Kalahari Desert.

Opistophthalmus holmi

Up to 60 mm long. A sand specialist, distributed throughout the central and southern parts of the Namib Desert dune system. It constructs burrows in consolidated vegetated sand dunes and may be found in or on vegetation at night. Because of its sandy coloration it is difficult to find without a UV light.

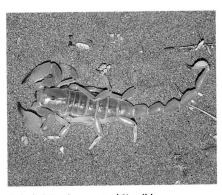

O. holmi, Sesriem, central Namibia.

SOUTHERN AFRICAN SPECIES

Opistophthalmus fitzsimonsi

Up to 65 mm long. Individuals from the northern parts of the distribution are darker. Burrows 10–15 cm deep in fairly hard soils. Burrow entrances are usually on open ground, near the base of vegetation or at the side of rocks. It is often found around the edges of salt pans in Botswana.

O. fitzsimonsi, central Kalahari Game Reserve, Botswana.

Opistophthalmus wahlbergii

Up to 110 mm long. Often shiny. Its appearance shows much variation across its wide distribution (over all of Botswana and Namibia, except the Namib sand system). It burrows in consolidated sandy soils. Burrow entrances are usually in the open, rarely against rocks or vegetation. In soft substrate, the burrow may be over a metre long.

O. wahlbergii, Witsand Reserve, Northern Cape.

Opistophthalmus opinatus

Very variable in size depending on location. Has an elongated sting. In southern populations, burrows are usually constructed in open ground. In the northern parts of the distribution, burrow entrances are at the side of large stones. Burrows descend to a depth of 15–25 cm.

O. opinatus, Maltahöhe, southern Namibia.

Opistophthalmus adustus
Up to 110 mm long. Found in southern Namibia, its distribution clearly restricted by the Orange River. It constructs multidirectional burrows up to 75 cm deep in sandy soils.

O. adustus, Rosh Pinah, southern Namibia.

Opistophthalmus flavescens
Up to 95 mm long. Its body is shiny in appearance, the mesosoma generally very dark as compared with the carapace and appendages. It is found in the Namib sand system where it constructs burrows, 30–50 cm deep in consolidated vegetated sand dunes.

O. flavescens, Swakopmund, Namibia.

Opistophthalmus macer
Up to 115 mm long. Found in short burrows under large rocks in fynbos areas in the Cape. Occurs in the same areas as *Uroplectes lineatus*.

O. macer, Worcester, Western Cape.

Opistophthalmus granifrons
Varies in length throughout its distribution. This species constructs burrows in the open in firm, gritty soils. Its limited distribution is an indication that it is a habitat specialist.

O. granifrons, Springbok, Northern Cape.

Opistophthalmus austerus

Up to 70 mm long. Constructs burrows 20–30 cm deep at the base of rocks usually at the top of hills. Females are exceptionally sedentary and do not venture too far from their burrows. As the females grow in size, they enlarge their burrows.

O. austerus, Laingsburg, Western Cape.

Opistophthalmus pallipes

Up to 130 mm long. It is found in rocky environments, where it constructs a scrape or shallow burrow under rocks.

O. pallipes, Springbok, Northern Cape.

Opistophthalmus karrooensis

Up to 100 mm long. Very similar to *O. pallipes* in appearance, but can be distinguished by forking of the medial suture at the front of the carapace. In *O. pallipes* there is no fork. Males have very elongated pincers. Found in rocky habitats, where it lives in a scrape under rocks and in rock crevices.

O. karrooensis, Beaufort West, Western Cape.

Opistophthalmus longicauda

Up to 150 mm long. One of the most beautifully coloured southern African species. Found in rocky habitats, where it constructs a burrow in the open or under rocks. Its burrows are about 25–30 cm deep. It can inflict a painful sting and is ready to use its venom in self-defence. This large scorpion stridulates loudly.

O. longicauda, **Upington, Northern Cape.**

Opistophthalmus capensis

Up to 130 mm long. Constructs a burrow under rocks in rugged areas. It is common on the Cape Peninsula. This species looks similar to *O. pugnax* which occurs to the north, in Gauteng.

O. capensis, **Langebaan, Western Cape.**

Opistophthalmus lawrenci

Up to 120 mm long. Localized distribution, confined to areas of red sand at the base of the northern slopes of the Soutpansberg, Limpopo Province. Constructs a burrow about 12 cm deep and 30 cm long. Pincers are quite smoothly textured. This is a relatively new species, previously described as *O. carinatus*.

O. lawrenci male (right) and female (left), northern slopes, Soutpansberg.

SOUTHERN AFRICAN SPECIES

Opistophthalmus pictus

Very variable in length, with a characteristic light area at front of the carapace. It constructs burrows in the open in very hard substrate. This species is very common in parts of its distribution and is one of the few in the genus that is found in the Free State.

O. pictus male (top) and female (bottom) showing the classic sexual dimorphism of the genus. Bloemfontein, Free State.

DISTINGUISHING FAMILIES BY FEET

OPISTOPHTHALMUS WAHLBERGII

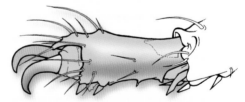

HADOGENES GRANULATUS

The lobed tarsus (foot) is characteristic of Scorpionidae. Stiff hairs on the tarsus help the scorpion to get a firm footing on the substrate.

This tarsal arrangement is characteristic of Ischnuridae. Highly recurved tarsal claws enable these rock-living scorpions to cling upside down to even the smoothest rock.

The shape of the pincers and sting is a good way to differentiate between genera.

		Sting	Pincer
BOTHURIDAE	Lisposoma	Shape varies. Often with subaculeur tubercle. Smooth in texture.	
BUTHIDAE	Afroisometrus		
	Uroplectes	*U. TRIANGULIFER* *U. PLANIMANUS*	
	Karasbergia		
	Lychas		Small in comparison with tail and sting. Smooth in texture.
	Pseudolychas	Short aculeus with subaculeur tubercle as above. *P. PEGLERI*	*PARABUTHUS TRANSVAALICUS*
	Parabuthus	Very round and rough in texture. *P. TRANSVAALICUS*	
	Hottentotta		
ISCHNURIDAE	Opistacanthus	Similar to sting of *Hadogenes*. 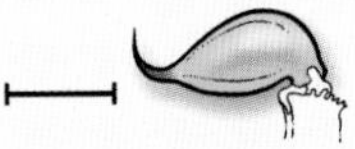 *C JONESII*	Roughly textured. 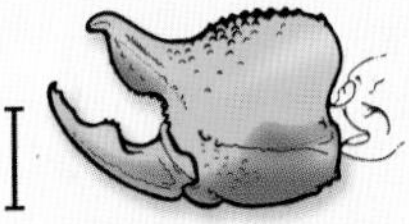 *C. JONESII*
	Cheloctonus		
	Hadogenes	Flat in profile. Short aculeus. 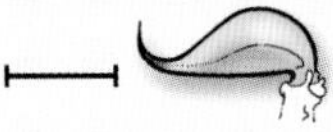 *H. GRANULATUS*	Powerful. No teeth on inside margin. 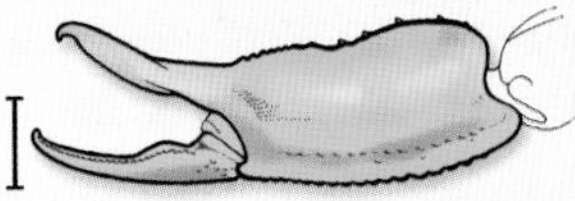 *H. GRANULATUS*
SCORPIONIDAE	Opistophthalmus	Sting elongated. 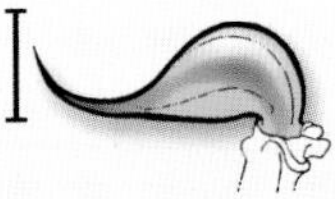 *O. WAHLBERGII*	Teeth on inside margin. Smooth in texture. 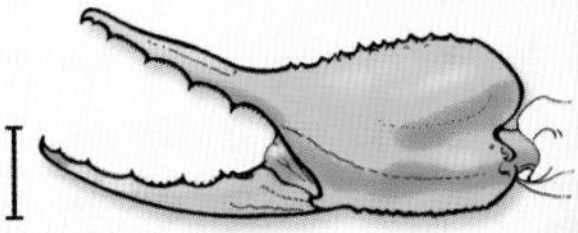 *O. WAHLBERGII*

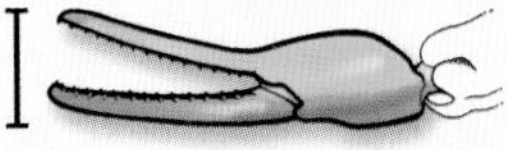

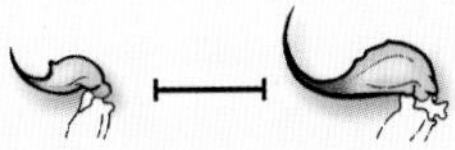

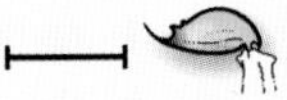

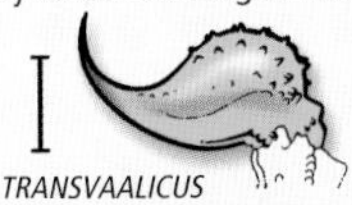

SCALE: Rule in each instance represents 5 mm.

LIVING WITH SCORPIONS

*scorpions do
not need to
live in close
relation
to man*

CONTROLLING SCORPIONS IS DIFFERENT from managing other problem invertebrates in that scorpions do not need to live close to man and other mammals or to live off food in areas of human habitation. The problem they pose for humans is simply that of incidental interaction.

In southern Africa, scorpions do not generally wander into houses although one species (*Pseudolychas pegleri*) has been known to enter dwellings in search of water. One other species (*Uroplectes triangulifer*) has been found inside houses, and certain arboreal members of the genus may seek shelter in thatched roofs.

In countries such as Mexico where scorpions are a major hazard, house construction plays an important role in scorpion control. Wire screens on all air vents, doors and windows, raised floors, smooth tiling, well-fitting roof tiles, close-fitting auto doors and filled-in wall cracks all help to keep scorpions at bay. In addition, the removal of surface debris, including rubble and vegetation, eliminates a favourable habitat for many Buthidae species. Fencing of a small mesh size, especially at ground level, is also important.

When camping in areas of high scorpion density, it is advisable not to sleep on the ground. Inspect bedding, keep tents zipped up, shake out clothing and footwear before use. Clear the camp area of rocks, logs and other surface debris. In places like the Fish River Canyon in Namibia overnighting next to bushes may seem like a good idea. It becomes less inviting when you consider that many scorpions in the area live on or near vegetation.

Be aware that some species of scorpion seek shelter in firewood. *Uroplectes vittatus, U. otjimbinguensis* and *U. formosus* may be found under the bark of fallen logs. These small, often cryptically coloured, scorpions have a painful sting. In northern KwaZulu-Natal, *Opistacanthus asper* commonly occurs under the bark of fallen trees and logs. In Sodwana Bay, a popular spot with divers in South Africa, these scorpions have been known to shelter in dive equipment (especially boots) hung out to dry on trees or bushes inhabited by the species. Standard precautions, like checking clothing and bedding, should be observed in such places and in scorpion-infested areas such as the Lowveld region of South Africa where *U. olivaceus* is responsible for a number of envenomations. Wear sturdy closed shoes when walking at night. Nocturnal patrolling with a UV light quickly reveals offending individuals.

Cycling in the Namib Desert in mid-summer is a good way to encounter many specialized species.

High concentrations of contact insecticides such as DDT and BHC are effective against scorpions, but are environmentally unfriendly. Household aerosol insecticides work on small species but not on medium to large ones – unless you bludgeon the poor creatures to death with the can. Insecticides only add to the problem by making the scorpion aggressive. Better to capture the offender and release it elsewhere.

LIVING WITH SCORPIONS

HOW DANGEROUS ARE SCORPIONS?

The venom of the majority of scorpions is harmless to man. However, a few species are highly venomous. All medically important scorpions in southern Africa belong to the family Buthidae, the members of which are characterized by thick tails and weak pincers.

All scorpions have neurotoxic venom. This means that the venom affects parts of the nervous system such as breathing, circulation, muscular co-ordination and blood pressure. Unlike some spider bites, all scorpion stings are extremely painful at the site of the sting. Where you are stung, the depth to which the sting penetrates and the amount of venom injected all have a direct bearing on the severity of a scorpion's sting. Because scorpion venom is neurotoxic, it worsens any pre-existing medical condition that affects the nervous system.

When humans are stung it is mostly at dawn or early morning and may happen in many different ways: treading on the scorpion with unprotected feet and ankles; putting on clothes or shoes that contain scorpions; sleeping directly on the ground in areas of high scorpion density (e.g. the Richtersveld in South Africa); carrying objects such as rocks or firewood under which scorpions are sheltering; carelessly handling scorpions.

HOW TO HANDLE A SCORPION STING

Stings from mildly venomous scorpions cause localized pain and swelling, with little systemic reaction. The affected limb should be immobilized, and ice should be applied, if possible, to the site of the sting. The site of the sting should be cleaned. It should not be cut open.

If a person is stung by a highly venomous scorpion, extreme pain is immediate, followed by muscle pain and cramps and other systemic reactions such as uncontrollable limb movements, numbness and difficulty with breathing and swallowing. Respiratory failure is the greatest danger for victims. Anyone showing systemic reactions, especially children and the aged, should be hospitalized immediately.

Spider- and snake-bite antivenom must **never** be administered to treat a scorpion sting. Victims should avoid alcoholic beverages temporarily and the use of steroids and anti-histamines. Tetanus immunization is recommended.

Venom sprayed into the eyes (a handful of *Parabuthus* species are able to spray venom over a short distance) produces an intense burning sensation and may result in temporary blindness if the eyes are not washed out thoroughly with clean water or some other neutral liquid such as milk.

There is no shortage of old wives tales on what to do if you are stung by a scorpion. Suggestions include eating the offender, applying crushed scorpion to the site of envenomation or soaking scorpions in olive oil for a few months and applying the oil to the sting. A traditional remedy is to rub the bark of the Maroela tree over the site of the sting.

Opistophthalmus pallipes male in hand.

Scorpions are not vectors of disease in humans though individual scorpions may be infested with worms or have mites attached to them. In small numbers, these cause the host little harm. A few species of mites attach themselves to scorpions only to hitch a ride.

Some people are allergic to scorpion venom in much the same way as to bee stings. In such instances, the sting of even a mildly venomous scorpion can have a severe effect.

Experiments conducted on laboratory mice and other animals are not a clear indication of toxicity or the effect on humans. The only information we really have on the toxicity of scorpion venom is from records of actual envenomations. It should be remembered also that venom is highly variable across the distribution of a single species, individuals in one area often differing from those of the same species in another area.

A toxicity scale for southern African scorpion genera is given on page 35.

At risk

The severity of scorpion envenomation in humans, according to medical research, is influenced by:

- any medical condition that compromises the nervous system
- body mass of the victim
- strength of the venom
- amount of venom injected

Children are more susceptible than adults because of their small body size. Elderly people are also at greater risk because of pre-existing medical conditions.

Many scorpions – *Hadogenes*, for example – are completely harmless to humans.

COLLECTING SCORPIONS

Collecting scorpions from the wild, for research and other purposes, is not difficult, but must be undertaken judiciously. It is important to have some knowledge of the species being targeted in order to decide where to search and which technique to use. You must ensure that you possess the necessary permits to collect, transport and keep arachnids. In addition, when crossing national and international boundaries you may be required to have an import/export permit. In South Africa, importing and exporting scorpions is strictly controlled through the issuing of permits by the provincial governments.

Scorpions are most active during the warmer months of the year. In the Northern Cape it is said that when a gentle hot wind blows from the north, scorpions abound. We know that scorpion surface activity increases after the first rains of the season, but we still do not understand all the factors that influence mass emergence events.

Collecting is most fruitful on dark, still nights with little or no moon. Other hunters may also be on the prowl as scorpions are a major food source for many animals. As smaller, more inconspicuous, scorpions tend to avoid times when more dominant species are active, they can be collected in sub-optimal weather conditions. In the Namib and Kalahari deserts, early morning is a good time to track individuals by following their footprints in the sand. A rake used on sandy substrate under bushes and grass tufts in these arid areas, and in the Northern Cape, may yield good results.

Many species are habitat specialists. To find arboreal species, trees and logs should be inspected, and peeling the bark off dead trees or fallen logs is often very successful. However, such species are best collected with a UV light, which causes minimal habitat disturbance or destruction. Fallen logs should always be replaced in exactly the same position.

Digging for scorpions in desert sand, Witsand Nature Reserve, Northern Cape.

Rock turning is another good way of finding scorpions. It involves locating suitable rocks – as a rule those large enough to provide a constant microhabitat. Rocks that are too small or thin warm up in the hot sun and are not good scorpion shelters. Scorpions generally do not favour rocks that are deeply embedded in the ground, although there are exceptions (for example, *Lisposoma* species). Rocks should be slightly embedded in the ground, with a narrow space at certain points between the rock and the substrate. Flat rocks that are situated on top of rocks and rock exfoliations provide ideal shelters for rock-living scorpions such as *Opistacanthus leavipes*, *Uroplectes planimanus* and for *Hadogenes* species. Once again, these rocks provide a narrow space in which the scorpion can shelter.

When turning over such rocks, always stand behind the rock and lift one side away from you. Depending on the species, the scorpion may run out from underneath. (A venomous snake may also be part of the deal!) Under wet conditions, however,

An arboreal species, *Opistacanthus asper*, hunting from its shelter in bark.

scorpions may be negatively geotaxic and cling to the underside of rocks. It is vital to put the rock back exactly as it was found. Scorpions are only one of the many creatures that use rocks as shelters. Ants, spiders, snakes, lizards, geckos and others live there too. Rocks should always be replaced gently.

Rock scorpions are also found in cracks in rocks and rock exfoliations. A torch shone into these fissures is sufficient to show up the scorpion or evidence of it, including exuvia and prey remains. These fissures can be prised open with a crowbar, but this method is very destructive and should be employed only if absolutely necessary. In prime scorpion habitat, never put your fingers into unseen nooks and cracks where scorpions may reside.

Although it is possible to use one's hands to take hold of some species of scorpion, long forceps should generally be used. The ends of the forceps should be rubberized to provide a soft gripping surface and to protect the scorpion from injury.

In some cases, scorpions (notably *Opistophthalmus* species) construct burrows under rocks, the entrance visible at the side of the rock. These burrows should be carefully excavated so as not to harm the inhabitant. Always fill in any excavations you make and replace any objects you may have moved.

Pitfall traps

Pitfall traps are unproductive except in the case of scorpions that burrow, such as *Opistophthalmus* and some *Cheloctonus* species. Instead of excavating (and destroying) the entire burrow to collect a scorpion, a pitfall trap left overnight at the entrance achieves the desired end with minimal impact on the environment and little risk to the scorpion through accidental damage. The scorpion can be examined and returned unharmed to its original burrow.

Pitfall traps should be inspected early in the morning as the scorpion will quickly perish if exposed to hot sun. Overnight rainfall, which may fill the trap with water and drown the scorpion, is another danger. So too are inquisitive animals, such as mongooses and birds, which may inspect the trap and eat the occupant.

UV light

It is well known that scorpions fluoresce under ultraviolet light in the range of 320–400 mm. Ultraviolet light is reflected by the hyaline layer in the exoskeleton, resulting in an eerie greenish glow. Newly moulted scorpions do not fluoresce until the cuticle hardens. This indicates that the substance responsible for the fluorescence is secreted shortly after ecdysis and develops as part of the tanning process of the exoskeleton.

On dark nights, even small scorpions, such as *Karasbergia muthueni*, may be seen as far as 15 m away with the help of a portable UV lamp. In areas where scorpions are plentiful, it is not uncommon to see over 500 specimens in a single night. This kind of collecting turns up more males than females, especially in the warmer months when males are looking for potential mates.

The spermataphore that male scorpions deposit when mating is also fluorescent and can be collected in the light of a UV lamp.

It is still not understood what causes scorpions to fluoresce under ultraviolet light.

Scorpions may be preserved in a mixture of:

6 parts formalin
15 parts isopropyl or
n-propyl alcohol (99%)
1 part glacial acetic acid
30 parts distilled water

Specimens to be used in DNA analysis should be preserved in 100% ethanol and stored in a freezer. This retards DNA degradation.

Dried specimens are fragile and break easily. To relax specimens for drying, a relaxing dish may be made from a flat-bottomed airtight container, the bottom lined with moist cotton wool covered with blotting paper. Specimens are placed on the blotting paper in an open container such as a petri dish. A mixture of water and fungicide (which can be any household disinfectant) should be added to the cotton wool. Specimens may be left in the airtight container for more than a day. It may take longer, however, before larger specimens relax and their joints become pliable.

There are other ways to relax large arthropods like scorpions, including dropping them into near-boiling water (after killing them more humanely). Smaller specimens need be immersed for only a few

seconds. Larger specimens may take longer. Alternatively, specimens may be immersed in Barbers Fluid for 30 minutes. This fluid can also be applied directly to the joints to promote flexibility.

Barbers Fluid consists of:
1 000 ml ethanol (95%)
1 000 ml distilled water
375 ml ethyl acetate
125 ml benzene

One way of relaxing a large arthropod.

UV-light collecting is preferable for a number of reasons. Scorpions do not seem to be disturbed by it, making it possible to observe courtship, orientation, prey capture, feeding, mating and other behavioural patterns in the field without disturbing the scorpions and with minimal environmental damage. Unfortunately, hazards such as snakes, warthog holes, rhinos and thorn trees do not fluoresce.

Portable fluorescent lights sold for camping purposes are easily equipped with a UV tube. Since UV light at these wavelengths cannot penetrate transparent plastic or glass, it is best to remove the tube cover. UV light can damage the eyes, and protective glasses are recommended for prolonged use.

CODE OF ETHICS

To ensure sustainable collecting, a code of conduct has been proposed for the collection of arthropods in general.

- Collecting of any organism must be undertaken responsibly.
- Specimens should be killed as humanely as possible.
- Only specimens necessary for the purpose should be collected.
- Specimens required for identification only should be released.
- Endangered, rare or localized specimens should be collected with great restraint.
- A single locality should not be sampled year after year.
- Indiscriminate collecting practices are discouraged.
- Unwanted specimens must be offered to other researchers for their collections, not discarded.
- Habitat disturbance should be minimal. Pitfall traps and excavated burrows must be filled in.
- Permission from the landowner must be obtained when collecting on private land. Permits must be obtained from the local provincial authorities where necessary.
- Specimens should not be collected for the manufacture of ornaments or jewellery.
- Collected specimens must be preserved and the relevant collection details noted.
- Collection data should be made available to science.

Every specimen that is collected for ecological, taxonomical and physiological purposes must be labelled with the following information:

- Name of specimen (if possible)
- Precise locality and (if possible) GPS co-ordinates
- Date of collection
- Name of collector
- Habitat details and collection method
- Reference number corresponding to that on notes and other data

Note: The length of a scorpion is measured from the tip of the carapace to the end of the tail.

Sustainable collecting is sometimes a matter of mutual tolerance.

SCORPION CONSERVATION

Habitat destruction and commercial over-utilization are just some of the threats to scorpions in the wild. Land in parts of southern Africa is being consumed at an alarming rate by urbanization and development, and it has become vital to protect our fauna and flora. Unfortunately, we still do not know enough about southern African scorpions to protect them fully.

Habitat destruction particularly affects species such as *Opistophthalmus lawrenci*, *O. ecristatus* and *O. jenseni*, which have very specialized habitat requirements and consequently are restricted to small areas. In Namibia, the known distribution of *O. intercedens* is about 40 km^2. *Hadogenes gunningi* is restricted to the Magaliesberg mountain range and discrete rocky outcrops. *H. gracilis* occurs nearby on a small number of rocky outcrops. *Hadogenes* species are heavily affected by commercial mining and quarrying activities.

Collecting for commercial purposes is often very detrimental to wild scorpion populations. Populations in the wild that are victims of indiscriminate and ignorant collecting may take many years to recover, depending on the species. This threat is very real in southern Africa, where many species have meek temperaments and weak venom.

It must be noted that the majority of southern African species are not suited to captivity. Many species are extreme habitat specialists and do not live long in abnormal conditions. Some species, meanwhile, are not suitable for captivity because they are highly venomous.

Mining and quarrying in prime scorpion habitat are among the many threats to scorpion diversity.

Scorpion biodiversity

Southern Africa is one of the most biologically diverse areas in the world and contains an exceptionally large number of plant and animal species relative to its land area. The rich scorpion fauna of South Africa and Namibia bears this out.

NUMBER OF SOUTHERN AFRICAN SPECIES BY COUNTRY

Family	Genus	Number of species						
		South Africa	Botswana	Lesotho	Swaziland	Namibia	Zimbabwe	Mozambique
BOTHURIDAE	*Lisposoma*	0	0	0	0	2	0	0
BUTHIDAE	*Afroisometrus*	0	0	0	0	0	1	0
	Hottentotta	2	1	0	0	2	1	1
	Lychas	1	0	0	0	0	1	1
	Pseudolychas	3	1	0	0	0	0	1
	Parabuthus	16	6	0	1	13	6	3
	Karasbergia	1	0	0	0	1	0	0
	Uroplectes	13	6	1	4	10	7	5
ISCHNURIDAE	*Opistacanthus*	6	1	3	2	0	2	3
	Cheloctonus	4	0	1	0	0	1	1
	Hadogenes	10	2	0	1	4	2	2
SCORPIONIDAE	*Opistophthalmus*	37	6	2	0	27	4	1
Total number of species		**93**	**23**	**7**	**8**	**59**	**25**	**18**

Glossary

aculeus: hollow, needle-like sting or stinger, used for injecting venom

booklungs: paired respiratory organs found on the underside of the mesosoma

carapace: shield-like plate of exoskeleton that covers the top of the prosoma

caudal segment: one of the five segments of the metasoma

chelicerae: pair of pincer-like appendages arising from the prosoma, used for feeding and grooming

cladistics: the classifying of animals and plants according to their genealogical relationship with other plants or animals

exoskeleton: hard supporting outer covering characteristic of arthropods

exuvia: shed skin that is discarded after moulting

fixed finger: fixed part of pincer that opposes the moveable pedipalp/pincer finger

instar: period of time between moults

keel: raised ridge-like projection of the exoskeleton

lateral eyes: two or three pairs of eyes situated at the front corners of the carapace

median eyes: pair of eyes situated near the centre of the carapace

mesosoma: mid-body region covered on the upperside by tergites and on the underside by sternites

metasoma: tail, consisting of five segments and the sting; adjoins the mesosoma

moveable finger: moveable part of pincer

opisthosoma: part of body consisting of the mesosoma and metasoma

pectinal tooth: tooth-like projection of the pectines

pectines: comb-like structures on underside of mesosoma

pedipalp: pair of three-segmented appendages arising from the prosoma, ending in the pincer

pincer: grasping part of the pedipalp, known as the hand

pleural membrane: thin flexible membrane between tergites and sternites on the sides of the mesosoma

scrape: shallow depression under surface debris

slit sense organ: tiny sensory slit in the exoskeleton used to detect stress in the exoskeleton

specialization: an adaptation that is specific to a species

spiracle: opening on the exoskeleton of the mesosoma for passage of air to and from the booklungs

sternite: shield-like plate of exoskeleton on the underside of the mesosoma

stridulate: to produce sound

subaculear tubercle: small tooth-like projection on inner curve of the aculeus

substrate: non-living material such as sand, rock or gravel in which an animal or plant lives or grows

tarsal claw: claw at the end of the tarsus

tarsus: foot, last jointed section of the leg

telson: end of the metasoma, known as the sting

tergite: shield-like plate of exoskeleton on upperside of the mesosoma

venom vesicle: part of the telson in which venom is stored

Bibliography

Bergman, NJ. 1997. Scorpion stings in Zimbabwe. *South African Medical Journal* 87: 163–167.

Eastwood, EB. 1977. Notes on the scorpion fauna of the Cape. Part 1 Description of neo-type of *Opistophthalmus capensis* (Herbst) and remarks on the *O. capensis* and *O. granifrons* Pocock species-group (Arachnida, Scorpionida, Scorpionidae). *Annals of the South African Museum* 72(11): 211–226.

Fitzpatrick, MJ. 1994. A new species of *Lychas* CL Koch 1845 from Zimbabwe (Scorpionida: Buthidae). *Arnoldia Zimbabwe* 10(3): 23–28.

Fitzpatrick, MJ. 1996. The genus *Uroplectes* Peters 1861 in Zimbabwe (Scorpiones, Buthidae). *Arnoldia Zimbabwe* 10(7): 47–70.

Francke, OF. 1982. Are there any Bothurids (Arachnida, Scorpiones) in southern Africa? *Journal of Arachnology* 10: 35–39.

Hewitt, J. 1918. A survey of the scorpion fauna of South Africa. *Transactions of the Royal Society of South Africa* 6: 89–192.

Hewitt, J. 1925. Facts and theories on the distribution of scorpions in South Africa. *Transvaal Royal Society of South Africa* 12(4): 249–276.

Lamoral, B & Reynders, S. 1973. A catalogue of the scorpions described from the Ethiopian faunal region. *Annals of the Natal Museum* 22(2): 489–576.

Lamoral, BH. 1979. The scorpions of Namibia (Arachnida: Scorpionida). *Annals of the Natal Museum* 23(3): 497–784.

Lawrence, RF. 1955. Solifugae, scorpions and pedipalpi, with checklists and keys to South African families, genera and species. Results of the Lund University Expedition in 1950–1951. *South African Animal Life* 1: 152–262.

Müller, GJ. 1992. Scorpionism in South Africa. *South African Medical Journal* 83: 405–411.

Newlands, G. 1972. Ecological adaptions of Kruger National Park scorpionids (Arachnida, Scorpionides). *Koedoe* 15: 37–48.

Newlands, G & Cantrell, AC. 1985. A re-appraisal of the Rock Scorpions (Scorpionidae, *Hadogenes*). *Koedoe* 28: 35–45.

Polis, GA (ed.).1990. *The biology of scorpions.* Stanford University Press, Stanford, California.

Prendini, L. 2001. A review of synonyms and subspecies in the genus *Opistophthalmus* CL Koch (Scorpiones, Scorpionidae). *African Entomology* 9(1): 17–48.

Prendini, L. 2001. Phylogeny of *Parabuthus* (Scorpiones, Buthidae). *Zoologica Scripta* 30: 13–35.

Prendini, L. 2001. Two new species of *Hadogenes* (Scorpiones, Ischnuridae) from South Africa, with a redescription of *Hadogenes bicolor* and a discussion on the phylogenetic position of *Hadogenes*. *Journal of Arachnology* 29: 146–172.

Index

Page references in *italics* indicate
photographs, tables or diagrams.